FEAR, WONDER, AND SCIENCE IN THE NEW AGE OF REPRODUCTIVE BIOTECHNOLOGY

SCOTT GILBERT AND
CLARA PINTO-CORREIA

FEAR, WONDER, AND SCIENCE IN THE NEW AGE OF REPRODUCTIVE BIOTECHNOLOGY

FOREWORD BY DONNA HARAWAY

COLUMBIA UNIVERSITY PRESS
NEW YORK

Columbia University Press
Publishers Since 1893
New York Chichester, West Sussex
cup.columbia.edu

Library of Congress Cataloging-in-Publication Data
Names: Gilbert, Scott F., 1949– author. | Correia, Clara Pinto, author.
Title: Fear, wonder, and science in the new age of reproductive biotechnology /
Scott Gilbert and Clara Pinto-Correia ; foreword by Donna Haraway.
Description: New York : Columbia University Press, [2017] |
Includes bibliographical references and index.
Identifiers: LCCN 2017000477 | ISBN 9780231170949 (cloth : alk. paper) |
ISBN 9780231544580 (ebook)
Subjects: LCSH: Human reproductive technology. |
Human genetics. | Biotechnology.
Classification: LCC RG133.5 .G557 2017 | DDC 618.1/7806—dc23
LC record available at https://lccn.loc.gov/2017000477

Cover design: Milenda Nan Ok Lee
Cover image: © Daniel Clowes. Courtesy of Fantagraphics

CONTENTS

PART V. EPILOGUES

FOREWORD

———

Making Babies, Making Kin

DONNA HARAWAY

This book in my hands, *Fear, Wonder, and Science in the New Age of Reproductive Technology*, like a human being, is itself fearfully and wonderfully made. Scott Gilbert and Clara Pinto-Correia come to readers as whole persons in this unusual and much-needed book. Their thinking, feeling, experiences, fears, curiosities, and hopes infuse these pages. Both developmental biologists, the authors engage a risky collaboration to present to ordinary people several stories that are usually kept rigidly separate. These accounts include (1) the rich story of how a human baby is made from the adventures of egg and sperm and the musical score of tissues and molecules to result in the emergence of a squirming infant from the body of a woman; (2) the nonlinear pathways in the history of developmental biology that bring us to our present scientific understandings and scientific narratives; (3) an account of technological, medical, and social strands that have been woven together in practices of biotechnologically assisted reproduction; (4) a respectful but pointed discussion of both the diversity of scientists' understandings of when human personhood may be said to emerge, and also of major world religions' diverse beliefs about when human life might be thought to begin; and (5) real-life stories of intense suffering, not only from involuntary infertility, but even more both from the pressure of enforced silence about one's infertility and from the misleading promises of a technical fix to a profound human experience. Each part of this rich tapestry of stories is woven in an acute consciousness of complex social, personal, and technical histories. Each part requires—as well as nurtures—emotional, intellectual, and sociohistorical intelligence. Each part is also alert to the demands of the shape and structure of narratives and images, the power of popular genres, and the pitfalls of false or just plain bad stories that cause personal pain and public misunderstanding.

Gilbert and Pinto-Correia's book is written for ordinary people, women and men of any age who care about how babies are made, how new sorts of kinship are invented, how biotechnology works, and how scientific, religious, and popular understandings change. The parts of this diverse book resonate with each other and draw me in as a reader. But still, at first, I was confused by the juxtaposition of stories and questions. Why was a striking account of the biological development of a human infant, along with the histories of biology crucial to understanding human reproduction in biomedical societies, in the same book with such potent accounts of the pain of involuntary infertility and of the technologies offered—and sold—to circumvent what is called many times in this book a "curse," and more, a curse crossing time and space to embrace all human cultures? Wouldn't readers dealing with involuntary infertility just suffer more reading about the bumptious adventures of human embryos in their rich, multi-actor worlds? What really ties the curiosity and awe inherent in developmental biology to the eloquent story of emotional, financial, cultural, and bodily suffering by women who want a bio-baby but cannot have one?

I think there are two potent answers to these questions, answers that make me love this book all the more. First, stories matter; they make and break human beings. And some stories are much better than others. Some stories are more true and therefore more able to sustain the fear, wonder, and science of human beings making family, making kin, with each other. Second, never before has it been so important to juxtapose the terrible realities of suffering and blasted futures from both too few babies and too many babies with the booming silences about each that prevent contemporary women and men from becoming able to respond to both. Finally, it should be no surprise that the first and second replies are profoundly related to each other.

First, the stories. Readers of this book belong to cultures that glorify heroic and competitive stories of individuals, economic success, and grand blueprints for future achievements. These accounts are usually full of masculine, privileged, usually white actors. Almost like a venereal disease, these stories have infected the generative organs of the scientific narratives of human reproduction, from DNA as the master plan of living beings to the exclusion of other formative actors, to monetized metaphors for everything on the planet, to fertilization as a contest rather than a collaboration, to the bounded (and boundless) individualism of everybody, including fetuses imagined to develop in solitary splendor. These same narrative genres have infected goals and means of technological development, lending power to the search for a technological fix to everything from climate change to reproductive dilemmas. *Fear, Wonder, and Science* disables these kinds of plots, offering instead a rich musical score for many players and many ways of making flourishing worlds for living and dying together on a mortal, bounteous, and finite earth. The conventional but still potent stories are disabled both by richer sciences and by better figural and narrative ways to conduct and express laboratory practices and knowledge making.

Similarly, the enforced shame of involuntary infertility also frays with better stories. That secret corrosive shame makes what might remain very hard for many people into something altogether different; namely, a life-long curse for women that blocks all sorts of generative responses, fulfilling lives, and ways of making community. The story and the experience of a "curse" of infertility must be undone, and this book proposes how. Those stories include the importance of listening to women experiencing painful realities, of refusing false promises, of celebrating lives that find support for suffering and the courage and cultural, social means to build lives without one's own children, not as second best, but as vibrant and needed personally and collectively. The needed stories also include the emphatic rejection of the false belief that there is an immutable biological imperative to the desire for and the right to have a bio-baby of one's own, no matter the costs. That bad story is like the tale of all-powerful DNA and self-acting fetuses. There is no evil curse, but there is widespread real suffering that requires a shared response emotionally, politically, and scientifically. Recognition of such suffering need not affirm the bad stories of biological reproductive imperatives. Recognition, with support, frees creativity and energy for women and families to flourish in other ways, including ways desperately needed on an earth with more than seven billion human beings in the grip of cascading ecological and social urgencies. That seven billion is now expected to reach more than eleven billion by the end of the twenty-first century—if birth rates remain low, as they are currently in most places.

Second, the suffering and the booming silences. Recognition of the complexity of involuntary infertility cracks the oppressive silences that break individuals and families. Also, better stories about developing fetuses interrupt the noise of conventional narratives that silence and dampen both awe and curiosity. Bad scientific stories cause actual suffering for real people, who are misled into believing determinist, competitive, and relentlessly productionist accounts of living and dying as mortal beings, as if those were the findings of good science when they are not. But the epilogues in *Fear Wonder, and Science* go much further.

These epilogues place the hard experiences of involuntary infertility and the rich stories of human development, including reproductive biotechnologies, together in a resolute rejection of determinism and the resolute courage to propose that making kin otherwise is the task that we must face together now, in fear and wonder, on an overburdened earth. A baby remains fearfully and wonderfully made, to be cherished in every way, and the experience of involuntary infertility remains fearful, to be brought into the open with good heart and cultural support. That needed support includes exposing the burden of infertility that results from environmental injustice and differential poisoning of whole regions, classes, and races, plus the dearth of reproductive health services for the poor, except for population-control-oriented contraception and sterilization. *Fear, Wonder, and Science* insists on making the unequal burden of suffering from infertility and from damaged lands and bodies central to breaking the booming silences that perpetuate reproductive injustice. Only then can the complex questions of both too few and too many babies be honestly addressed, in fear and wonder, with the help of sciences in non-arrogant alliance with caring people coping with unprecedented revolutions in kinship.

The shared need is for imaginations and practices for making lives, including making kin who are not necessarily biologically related. Celebrating the birth of a baby remains a good thing, and the intricacies of biological development are no less full of wonder. But learning to celebrate and proliferate practices—emotional, intellectual, technical, and cultural—for making kin that are not tied to making babies is urgent.

The booming silence among just about everybody except population professionals about the tremendous increase of human numbers, and the demands those numbers make, especially by the rich, is broken in this book. The silence is replaced by a vibrant musical composition embracing babies, sciences, women, men, bodies, stories, and possibilities. Perhaps, at the end, *Fear, Wonder, and Science* is less about assisted reproductive technologies and more about assisted kin-making practices. The latter include the former—assisted kin-making practices include assisted reproductive technologies—but enfold reproduction into quite another story of human beings learning to flourish with each other and other mortal critters. That is fearful and wonderful, indeed.

PREFACE

A biologist, a theologian, and a philosopher walk into a bar. They are very happy to do so, since the Finnish winter night is chilly, and there's an eager crowd waiting for them inside. Seated on a makeshift stage, a pert graduate student asks them, "What is the single most important story in the world?"

The theologian responded immediately, "Salvation through God's grace." He goes on to expound the mysteries of the Cross.

The philosopher looks disdainful and replies, "The Enlightenment." He talks about the life of the mind and the discovery of truths.

The biologist feels that he is expected to say "Evolution." But that isn't the most important story. In fact, that is a consequence. "The most important story in the world," he says, "is the construction of the embryo."

That is why we wrote this book. We are, as Psalm 139 proclaims, "wonderfully and fearfully made." Indeed, there are very few people who can even *imagine* how wonderfully and fearfully made we are than developmental biologists, those scientists who have the privilege of studying embryos. And the two of us have the good fortune to be developmental biologists. Although we disagree on many things, we agree on the awe, harmony, and mystery of the body. And we agree that these are stories that need to be told. We also agree that these are stories that are constantly being mistold to the public. These false stories degrade the human embryo. They also degrade the human ingenuity of the biological sciences that is discovering how fertilization occurs, how we are constructed from the one-celled embryo created by this event, and how this knowledge can be used for human happiness. For us, hearing politicians, theologians, scientists, and media commentators spout absurdities about human embryos is nothing less than hearing one's beloved being vilified. In addition, we are both angry when change is uncritically touted as progress without investigating the possible damage such change can make to real people and their families. Therefore—and not without much arguing—we agreed that we should put some of these ideas about reproductive science and its related technologies into a book.

From the start, we've been trying to write a book with potential benefits for several different audiences: parents and partners, professors and students, teenagers seeking knowledge forbidden to them by their schools, young women who are not infertile but voluntarily postponing childbearing, clinicians applying their skills to infertile patients, and the struggling, hopeful patients, themselves.

Also, the story of the embryo matters to all human beings. Whether we're aware of it or not, how our bodies are built forms the first part of our autobiography. People want to know at what stage human life begins. People want to know how fertilization takes place and when should one use in vitro fertilization. People want to know what stem cells are and how they might be used to alter human longevity, health, and welfare. We figure we can tell these stories.

We are very similar and very different from each other. Both of us have advanced degrees in biology and the history of biology. Holding degrees in the humanities makes us different from most scientists. In terms of our integrating science with social events, Clara was a student of Stephen J. Gould, and Scott was a student of Donna Haraway. Both of us got excited about developmental biology—the science of how the embryo comes into being and makes its different organs—and, as research scientists, we investigated these processes while maintaining a creative life outside our laboratories. Clara is a renowned novelist and media commentator; Scott plays piano in a klezmer band and has written several biology textbooks, as well as articles on art and the Bible.

But we are also very different in our values and social contexts. Clara is a Portuguese Catholic woman whose socialist views can be seen in her novels. She also has firsthand experience with infertility clinics in both the United States and Europe. Scott is a Jewish man from New York, whose wife is an obstetrician–gynecologist with whom he has three children.

Scott has had a stable position at Swarthmore College for over thirty years. Clara has experienced the economic turmoil and academic calamities of southern Europe.

Therefore, we have different voices and different opinions on the many issues we shall discuss here. To make the most of our backgrounds and strengths, we separated our respective jobs as much as we could.

Clara and Scott will be writing separate, but intertwining, chapters. Scott's chapters will tell you, step by step, how the marvelous adventure of life begins: how sperm and egg cooperate in unexpected ways such that these two cells, each on the precipice of death, can generate the foundations of a new organism; how organs help each other to form; how twins form; and how different types of gonads arise.

Clara's chapters provide the human counterpoint to the science discussed in Scott's chapters. How do these scientific discoveries affect real people? She will tell you stories of how assisted reproductive technologies entered human lives in the mid-1970s, describe the speed with which they have been adopted, and look at how the commercialized versions of these techniques expanded the market from women struggling to end their infertility to women choosing an alternative lifestyle. These are the stories of people who desire a biologically related child, and who are increasingly seeking assisted reproductive technologies as their only hope to build such a family. And it is also a story of the emotional and physical costs of these procedures.

Do we have answers to everything? No, of course we don't. Scientifically, though, we have data that fit certain conclusions better than others. But mostly, we have great stories. We have the story of how the cells of your body came into existence, how you developed testes or ovaries (and sometimes, but rarely, both), why you have only two eyes (that are always in the head), and other great mysteries. We also tell a story of how scientists figured out how to enable infertile couples conceive and bear children. This is a story of biblical proportions. But it is not a story of unalloyed successes. Rather, it is a story of incredible joys, incredible sorrows, and incredible courage. In short, we are telling the stories of life in the twenty-first century.

Back to the Finnish bar:
L'chaim! To life!
Saude! To your health!

ACKNOWLEDGMENTS

This book is at the junction of two fast-flowing streams. One tributary is the 2005 volume *Bioethics and the New Embryology*, coauthored by Scott, Anna Tyler, and Emily Zackin, published by Sinauer Associates. Andy Sinauer and his staff have been wonderfully generous in allowing us to use some of the material and illustrations from that volume. The other tributary flows from a book-long interview that Clara had published concerning the human toll of in vitro fertilization failure. What you see here is a confluence of those streams. Some of the spirit of these original books is continued into this very new and different volume.

And it takes a village to write a book! This book has benefitted enormously from the knowledge and wisdom of an entire dispersed tribe of people. The form of this book was conceived by Scott and Clara in collaboration with Dr. Jason Robert (Arizona State University) at a meeting with editors Patrick Fitzgerald and Ryan Groendyk. Since this is an unusual publishing format, integrating contemporary science with firsthand narratives and sociology into signed chapters, we had to be very careful, but also very excited, about the prospects of this new type of book. We could not have gotten better editors for this experiment. We also wish to thank Kara Stahl, whose editorial skill and judgments could not have been better for this project.

Clara wants to seize this occasion to thank those who made it possible for her to write the final version of her chapters simply because they were there when she needed someone to help her make it through times of distress: "If it weren't for my sisters' backing and the constant emotional support João and Ana gave me, I would not have been able to finish this project. For the same reasons, at different points along the line, I owe special thanks to Ana Maria and Paulo (I stayed at their house whenever I had to go back to Lisbon), to Marga and João and Cagijo (I worked at their restaurant Número Nove when my own internet was down), to Fernando (his place sheltered me through many an all-nighter), and finally to Madalena (I debated my ideas with her when I wasn't sure of the best strategy to tackle delicate issues and ended up benefiting greatly from her endless sources of online information)."

Scott wants to thank the people who were influences on this book. These certainly include our teachers, Dr. Barbara Migeon, Dr. Stephen J. Gould, James Robl, and Dr. Robert Auerbach. Who we are depends on whom we meet, and we acknowledge our inductions from these and other mentors. There were many people along the way, influencing us less formally. Certainly the discussions with Anne Fausto-Sterling, O of Serenity House, Philadelphia and numerous students about teaching social responsibility in science courses has been incredibly important.

We sincerely thank Dr. Donna Haraway, who was one of Scott's teachers, for her foreword to this volume. She has been the inspiration not only for much of this volume but also for so much of contemporary science studies. And a major power in the gestation and labor of this book has been Scott's wife Dr. Anne Raunio, a clinical obstetrician–gynecologist who has intimate knowledge of the wonder and fear of human development and birth. Not only did she contribute data and ideas, she also acted as our first editor, making certain that our information was accurate and readable. Without her, this book could never have come into being.

We also needed reviewers who were sympathetic to the goals of the project and at the same time critically astute. Friends don't let friends publish garbage. These friends and colleagues include developmental biologists Dr. Rocky Tuan (University of Pittsburgh), Dr. Dominic Poccia (Amherst College), Dr. Susan Squier (Pennsylvania State University), and Dr. Anna Edlund (Lafayette College). Each has a different specialty, Rocky being a stem cell expert, Dick concentrating on fertilization physiology, Susan focusing on medical reporting, and Anna teaching courses in the biology of women. A huge thank-you goes to Dr. Katayoun Chamany (The New School, New York), who has expertise in each of these areas and who also teaches courses in science and social justice.

And we thank the numerous people who, when they heard of our project, supported us and told us that we had to complete it no matter what. We sincerely hope this book doesn't disappoint any of them.

Scott Gilbert
and Clara Pinto-Correia

FEAR, WONDER, AND SCIENCE IN THE NEW AGE OF REPRODUCTIVE BIOTECHNOLOGY

———

I

THE IMPORTANCE
OF THE STORY

H umans are the storytelling species. This sets *Homo sapiens* apart from all other animals and is a common denominator across all human cultures. Every human social group likes to tell and hear a good story. And we tell ourselves stories all the time.

We will be presenting some amazing stories in this book. One of them is the story of fertilization. Another is the story of embryonic development. Still another story concerns the triumph and anguish of infertility treatments. These stories are intimate tales of desire, cooperation, exhilaration, and despair; and they are linked together in unexpected ways.

So, our first duty is to look at how we tell our narratives. How do we construct our stories? Using the notion that we need "defense against the dark arts," chapter 1 seeks to give the reader tools that will enable him or her to rationally consider the information about who we are and how we came to be. It looks at how science is described and how those descriptions can alter the way we view the topic. Language and images can make us accept unscientific beliefs, even when they are presented as science.

Chapter 2 looks at the narratives of assisted reproduction. The stories of assisted reproductive technologies are often told as conquests over adversity, if not conquests over nature itself. But for many, assisted reproductive technologies have been a double curse: first, as a result of the infertility, which is seen as a curse in all cultures; and second, as a curse upon their lives, since the largely unregulated economics of the field allow some people to prosper from the misfortunes of others. This chapter details both types of curses.

1

CONCEPTUAL DETOX

———

Returning to Hogwarts to Learn Human Embryology

SCOTT GILBERT

Curiosity is not a sin. . . . But we should exercise caution with our curiosity . . . yes, indeed.

—Albus Dumbledore, *Harry Potter and the Goblet of Fire*

THE DEFENSE AGAINST THE DARK ARTS

"I teach developmental biology, but for the next three weeks, I am your instructor in the defense against the dark arts."

This is how I introduce myself to the introductory biology class. The students laugh, of course, understanding my reference to some of the strangest faculty of the Hogwarts School of Witchcraft and Wizardry. But I challenge them: "Why are you laughing? I'm serious. There are people casting spells on you to make you believe things that are not true and to vote or spend your money based on those misbeliefs. Could you defend yourself against a person who claims that your attitudes and personality are determined by the genes you receive at fertilization? Could you argue against someone who says that the fastest sperm gets to fertilize the egg and it does so by actively drilling a hole in it, or that morning-after pills cause abortions? None of those statements are true, though they are widely believed. There are people casting spells on you, telling you that these are scientifically valid truths and that you should trust them.

My job is to teach you the counter-spells, to teach you what biology actually knows about fertilization and early human development."

Indeed, we will be telling stories of some remarkable cells, starting with the sperm and the egg. But sometimes, this story does not occur. Sometimes the egg and sperm don't meet, or the embryo perishes. So, after discussing normal human development, we'll be discussing why fertilization sometimes cannot take place and how assisted reproductive technologies (ART) can often circumvent such issues. This is a story of technology, human creativity, and how we've used our knowledge of fertilization to make therapies that have the potential to help people. Like the story of the embryo, this story is also a heroic tale, but that of human, rather than natural, creativity. However, the public is often misinformed about the natural process of fertilization and our technological abilities available to assist it. So, our first duty is to identify some spells and give you the charms to counter them. This means looking at language.

Most of what we learn about human fertilization and early development comes not from scientists, but from other wizards—filmmakers, theologians, cartoonists, and journalists. They use images, words, and stories to make us think that nature conforms to a particular set of values, when it often doesn't. They cast spells on us. Spells are made of language and images, which are the tools we use to construct our notion of what is real. Indeed, one of the major functions of biology is to enable people to understand things that can't be seen. Therefore, we often describe similarities between microscopic events and more familiar events. We write that the mitochondrion is the powerhouse of the cell or that cells are the bricks of the body. In this way, we convey the microscopic conclusions that mitochondria make energy for the cell and that the body is composed of cells.

Thus, we use language to understand what we can't see, saying that the unseen object—a cell, a sperm, a chromosome—is similar to something well known (Lakoff and Johnson 1980). There are four major ways of conveying these similarities: via similes, metaphors, analogies, and images. So, in order to counteract false spells, we need to learn the rules of language and images.

SIMILES AND METAPHORS

Somewhere in school, we were taught the difference between similes and metaphors, and we parroted the boring phrase that "similes use 'like' or 'as.'" Actually, few things in this world are more important than the difference between simile and metaphor. Both similes and metaphors concern similarities between two unlike things. But similes are intellectual comparisons, and they use "like" and "as" to establish rationally that the unknown object has a characteristic similar to the known object. Metaphors, however, make an equation of one thing to the other. Whereas similes are rational, metaphors are magical.

For example, in the Bob Seger song, "Like a Rock" (1986), the singer claims, "Like a rock, I was strong as I can be/Like a rock, nothing ever got to me." This man is not claiming to be a rock, but merely to have two of the properties of rocks: strength and endurance. However, in "I Am a Rock" (1965), when Paul Simon sings, "I am a rock/I am an island," he constructs an emotional image, a magical claim to be this rock, this island apart. Similarly, when Oscar Hammerstein Jr., wrote in "All the Things You Are" (1939), "You are the promised breath of springtime," he was making an emotional equation that had nothing to do with his beloved's similarity to the properties of delayed exhalation. Similes are rational; metaphors are emotional. Similes tend toward the intellectual; metaphors tend toward the magical, subverting intellectual discourse.

Metaphors and similes are incredibly important in explaining things that can't be seen. That's why scientists use them all the time. If I say that amino acids are the building blocks of proteins, you can get the idea that a protein is made from constituent parts called amino acids. If I say that a cell moves over a bone like a tractor, you can get a feeling for how that cell is moving, even if it doesn't have a motor or tires. Metaphors and similes make the unfamiliar familiar. Because of this, it is probably impossible to communicate biological science without metaphor or simile. Biological science could not exist without them. Therefore, we do not need to defend ourselves against metaphor or simile. We need to defend ourselves against their improper use. The improper use of similes and metaphors can give us a very distorted picture of nature, indeed, a picture at odds with what science actually knows.

Metaphors provide mental pictures, images, of microscopic reality. And very often, all that is needed is a word to put an idea into a particular framework. For instance, argument can be described using the metaphor of conflict. "Your claims are not defensible," "He attacked a weak point in my argument," "I shot down his ideas," et cetera. But argument can also be described using the metaphor of path: "His argument went nowhere," "He had a long argument, but it came close to the truth." Science uses such extended metaphors all the time (Lakoff and Johnson 1980; Gilbert 1979; BGSG et al. 1988). For instance, the early human embryo interacts with the mother's uterus, the womb. Picture these four sentences:

1. The embryo implants into the uterus.
2. The embryo docks onto the uterus.
3. The embryo burrows into the uterus.
4. The embryo invades the uterus.

The verb in each sentence is a metaphor for some visible process: planting, boating, digging, or combat. Each one describes some property of the interaction between the embryo and the uterus. The metaphor that's chosen will permit us to think in that particular way and not in the manner of the other metaphors. That's why they are so important. The metaphors we use to describe things channel our thoughts. And I just used a metaphor— "channel." It gives the image that of all possible ways of thinking, the metaphor chosen allows us to follow only one path.

AN EXAMPLE OF A POTENT METAPHOR: DNA AS SOUL

So let's look at one of the most powerful and false metaphors in the public's understanding of modern biology. One of the great spells being cast on us today is that deoxyribonucleic acid (DNA) is our soul. DNA, of course, is the molecule that forms the core of our genes. Our chromosomes are strings of genes composed of DNA and proteins. And if one changes the DNA of a gene, changes often result in the body. So DNA is a very important molecule. The idea that DNA is our essence runs incredibly deeply in American culture. My family receives a newsletter from the Finnish–American Society, in which one editorial claimed, "The sauna is in the DNA of every Finn." Of course, what the author meant was that "the sauna is in the *soul* of every Finn." But "DNA" has become the secular equivalent of the religious word "soul." This was documented in the 1980s by sociologist Dorothy Nelkin and historian Susan Lindee (2004), who sought to discover how DNA was represented in popular literature. Rather than read scientific literature, they read newspapers, *Newsweek*, *Time*, *Vogue*, *Redbook*, and other periodicals that you might find in your home, a doctor's office, a hairdresser's shop, or a train station. What they found was fascinating. DNA, the hereditary material of our genes, was depicted as if it were our very soul: (a) it was our essence; (b) it determined our behaviors; and (c) it could be used to resurrect our bodies (as in *Jurassic Park*.)

One surprising place to see how deeply entrenched this metaphor is in our culture is to look at automobile ads. On websites and in magazine ads, we learn that "superior handling is in the DNA of every German sport coupe" and that the new Kia Sportage is "genetically modified." An ad for the Jeep Compass claims that its "red-blooded attitude" is "in the genes." And, an ad for Infiniti notes, "While some luxury sedans just look like their elders, ours have the same DNA." Probably the best car ad showing the perceived link between DNA and essence is that for the midsized Hummer. An ad in the May 23, 2005, *Newsweek* shows the car over the tagline, "Same DNA. Smaller Chromosomes." In other words, the size may be smaller, but the essence hasn't changed. Biologically, this makes no sense. (I mean, cars don't have DNA. We know this.) But the ad informs us that although the size may be smaller, the vehicle has the essence of a real Hummer (Gilbert 2015a). DNA has become who and what we are.

And this DNA, we are told, controls our behaviors. Newspapers are full of stories proclaiming that scientists have found the genes for schizophrenia, bipolar disorder, homosexuality, musical ability, and sadness. We haven't. Often, the results of one study will be touted by the press, but when further studies are done, the first study is shown to be a statistical aberration. These further studies don't get reported. We don't hear that the gene for schizophrenia has been "lost." But the public latches on to stories about how our behaviors are determined by our genes.

Recently, several reputable newspapers reported that scientists found that mutations in a particular gene make one liberal or conservative (Gilbert 2015b). Fox News tagged this "the liberal gene" and told its listeners, "Don't hold liberals responsible for their opinion—they can't help themselves. A new study has concluded that ideology is not just a social thing; it's built into the DNA, borne along by a gene called *DRD4*." The *National Examiner* claimed that this research confirmed "Joy Behar's liberal birth defect."

What the cited paper (Settle et al. 2010) actually said was that "the *7R* allele of this gene has been associated with novelty-seeking behavior, which is a tendency that is related to openness, a psychological trait that has been associated with political liberalism." In other words, a gene variant has been *associated* with a trait (i.e., it has not been shown to cause it) that has a *tendency* (which means there are many exceptions) to be *associated* with (not causing or caused by!) another trait that has been associated with (not causing) political liberalism. This is not the best chain of causation, but one that was trumpeted widely throughout the United States. Another gene and its variant were similarly touted as making a person either a warrior or a worrier. It didn't turn out to be that way. Our behaviors are formed by genes, friends, economic conditions, parents, and a whole host of other factors. Yet television ads are telling us to send in our DNA samples to find out "who we are." DNA has become our soul.

This idea has incredibly important ramifications. But let's look at how this metaphor of DNA as soul plays out in the debates on abortion and stem cell research. Remember, we get our unique set of genes at fertilization. Thus, in this way of thinking, fertilization becomes the equivalent to the religious notion of "ensoulment"; when we get our soul, when we become a person (Gilbert 2008). According to this view, we receive our soul when we get our DNA; that is, at fertilization. No wonder numerous right-wing politicians in the United States have been saying that DNA confirms that our souls are given to us at conception (Gilbert 2015a). As we will see later, there are many other places where scientists think personhood arises; but the metaphor of DNA as soul, which is so pervasive in our society, predisposes people to think that fertilization is the time of ensoulment.

We can find this notion throughout the web and in many books and magazines (Nelkin and Lindee 2004; Gilbert 2008, 2015a). One theologian (Ramsey 1970) tells us, "Genetics teaches that we were from the very beginning what we essentially are in every cell and in every human attribute." Actually, genetics teaches no such thing. The genes permit one to develop in certain ways and prohibits development in other ways. They do not determine who we are. This idea of genetic determinism was a major point in the abortion debate of the 1970s (Greenhouse and Siegel 2012), and several very popular websites (e.g., justthefacts.org, mypregnancysolutions .com, prolifeinfo.ie/life/amazing-facts/) currently tell their readers, primarily teenage girls seeking to learn the facts of life, "Even more amazingly, intelligence and personality—the way you look and feel—were already in place in your genetic code. At the moment of conception you were essentially and uniquely you." This is also false. (Often these sites are run by anti-abortion agencies that do not reveal as much on the site.) But notice that the word "essentially" is used to describe the relationship between a person ("you") and his or her DNA ("your genetic code"). In both cases, this essence is made from the DNA. This view is not uncommon. DNA has become our essence.

But this belief that DNA is our essence, our soul, is metaphor. One way of combatting this spell is to convert the metaphor into a simile. This is the "Finite Incantatem" charm learned at Hogwarts. When one says, "DNA is like our soul," it gives one permission to ask, "How so?" Here, the metaphor fails. It is not the controller of our fate. Ask any parent of identical teenage twins (whose genes are identical) if these children are the "same." Rather, "who" we are is determined by a complex array of genetic biases and personal experiences, including (as in the Harry Potter books) parental affection.

Making metaphors into similes allows one to combat a false or partial metaphor with alternative metaphors and similes. To some scientists, DNA is like a library. To Natalie Angier (1992), DNA is like a powerful politician who is told to say different things in different cells. I often compare DNA to a musical score. We are each performances of that score, and every time one plays it, it is different. Identical twins would be different performances of the same score. Metaphors are magical devices used to control what you think. It is crucial to recognize them and to challenge them.

Now, as I have just mentioned, experience, even maternal affection, can alter DNA. This may sound like science fiction, but we have excellent scientific evidence for this. This is like the "Patronus" charm in the Harry Potter series. It directly defends one against mischievous spells. The evidence comes from the science of neurobiology (the study of the nervous system) and epigenetics (the study of how genes get activated in different types of cells). So how can maternal affection alter DNA?

Most of our laboratory animals have been inbred by mating them to their siblings for generations. Therefore, strains of laboratory rats and mice have the same genes as other mice of that strain (except for the ones that causes sex differences). Every rat of a particular strain is like an identical twin to any other rat in that strain. Yet some rats are anxious, and others are not; some rats give their infants abundant care, and others do not. If their genes are the same, what is causing the difference?

It turns out that a mother rat's affection—demonstrated by stroking and licking her newborn pups—generates hormones in the pups that go to the brain and act to remove methyl groups from the DNA of certain genes. This loss of methyl groups (one carbon and three hydrogen atoms, the same methyl as in methanol) from the DNA activates these genes, which include those involved in making the hormones that promote certain behaviors, such as calmness and caring for young. The mice that have had sufficient maternal care now have different DNA from those who did not get adequate maternal care, even though they started off with identical genes (Meaney 2001; Champagne et al. 2006). Those mice who did not get adequate maternal care when young generally become more anxious as adults, and they aren't as interested in nursing their own pups. Here, behavior (maternal care) alters the DNA! The inherited DNA is the same in both cases, but it gets modified by the environment. Sometimes, when we read magazines and hear advertisements saying that DNA is our soul or that we are slaves to our DNA (another metaphor!), we must remember that the activity of DNA can be modified by our environment and chant it to ourselves often.

ANALOGIES: THE LANGUAGE OF NATURE AND THE NATURE OF LANGUAGE

Analogies claim a similarity between relationships. For instance, "A puppy is to a dog as a kitten is to a cat." This could be written as follows: puppy:dog = kitten:cat. Shakespeare used metaphors to make one of the world's most famous analogies: "All the world's a stage and the men and women merely players." People are to the world as actors are to the stage (and each of us has his or her roles to play). Scientists use analogy all the time, and it is a major tool for popularizers of science. Indeed, analogies are commonplace. Think of such analogies as (1) food:body = fuel:car, or (2) cell:body = brick:building. Analogies show that the relationship between one pair of items is the same as the relationship between a second pair of items.

An Example of a Potent Analogy: Manly Sperm and Feminine Eggs

In the public's view of biology, one of the strongest analogies is sperm:man = egg:woman. Of course, sperm are made by men, and eggs are made by women, but the analogy goes deeper in the public mind. Here, the sperm act as men act, and the eggs are surrogates for women. This analogy became explicit soon after fertilization was discovered, when one of the leading textbooks claimed that the sperm were suitors, and the egg decided which one would enter her (BGSG 1988; Gilbert and Fausto-Sterling 2003). Here, fertilization was the marriage of sperm and egg. Indeed, one has to recall that the scientific term for sex cell (i.e., the sperm or egg) is "gamete," which means "marriage partner."

Later, the sperm and egg became depicted as characters in a self-congratulatory hero myth, in which the sperm undergo a long and dangerous odyssey, and the victor wins the egg as his possession. Indeed, one of the most popular books in this area tells us that "the sperm undergo a perilous journey," and "the successful sperm surround the prize." One version of this story reminds one of the "Sleeping Beauty" story. Here, the egg is dormant until wakened by the binding of the sperm that had undergone the quest to find it. In these later stories, the sperm is active, but the egg is simply passive but receptive to the sperm (Schatten and Schatten 1983). The traditional values of active men and passive women are thus described as being replicated in nature.

Recently, the stories have become more militaristic. As men become road warriors and real warriors, the sperm are also seen to be soldiers. One popular article depicts sperm as the ultimate warriors in the never-ending battle against the egg and against other sperm. Sperm are here described as "tactically smart," "well armed," and as "a formidable 0.00024-inch weapon, tipped with a chemical warhead" (Small 1991)! The egg is described as being both "fortified" and "sending out alluring chemical cues." Rather than being a Sleeping Beauty, the egg is now Helen of Troy. The cultural stereotypes of powerful, sexually attractive women and tactically smart warrior men are transferred to the sperm and egg. As we will see in chapter 3, cells don't act like men and women. And such stories are being used to make people think that culturally accepted gender roles have a biological basis in the ultimate sex partners—the sperm and the egg.

The Hogwarts charm to counter this type of spell is "Riddikulus"— making fun of it. We've just turned what appeared to be a scientific story into a fairy tale and a myth. It's been deflated and its power over us removed.

IMAGES

Whereas metaphors and analogies may create images in your mind, there are many images that come ready-made. We see them all the time in magazines, newspapers, and books and on websites. The first thing to realize is that there are no uninterpreted cells and no uninterpreted fetuses (Gilbert and Braukmann 2011). These things are hidden from our view, and when we make them apparent, even by photography, we change them.

An Example of a Potent Image: The Human Fetus

Images have incredibly important consequences for determining our view of embryos, for very few people have ever seen an actual human embryo (Gilbert and Howes-Mischel 2004). When we think of human embryos, we are told (by magazines, websites, and placards) to think of a living entity with arms, legs, fingers, toes, and eyes. Similarly, the adjective "embryonic" connotes such a formed being. But this image is actually not that of a human embryo; it is of a human fetus. When biologists talk about human embryos, they are talking about early development (the first eight weeks of gestation) (figure 1.1). Embryonic stem cells, for instance, are not derived from an organism with a head, limbs, or torso. They are derived from a cluster of cells that have neither front nor back, left nor right, head nor belly. While articles about embryonic stem cells are often illustrated with pictures of ten-week-old fetuses, embryonic stem cells are derived from a nondescript ball of about 100 cells (about eight days old) that have no observable structure. The embryo at this stage is called a blastocyst (a fluid-filled ball of dividing cells). Its outer ring forms part of the placenta (the chorion), whereas the central ball of cells are the "embryonic stem cells." They are called "embryonic" stem cells because they will make the embryo (whereas the outer cells make the placenta). As you can see from the figure, they have no head, arms, or legs.

There are other images of fetuses that have become popular. One was a picture of a human fetus on the June 9, 2003 cover of *Newsweek*. The caption reads, "Should a fetus have rights? How science is changing the debate." But the fetus on the cover is not anything found in nature. First, it is free-floating and not connected to a mother. Second, it has no placenta. Third, it has no umbilical cord! The fetus was presented as an independent organism, something it definitely is not. The caption should have read, "Should a fetus have rights? How Photoshop is changing the debate." But the free-floating fetus has become an image in our mind. Like a fairy, gnome, or unicorn, it does not exist. It is a spell.

Indeed, these images of embryos were the important part of the article—because they convey an idea of what you should be thinking. As the Pro-Life Action League (2003) exulted:

> Some complained that the text was not very pro-life, but the photographs and computerized pictures were worth tens of thousands of words, and there were twenty-four such pictures in all. We were astonished at the power of this presentation by an avowedly not pro-life publication. We bought extra copies and showed it to everyone.

Images matter.

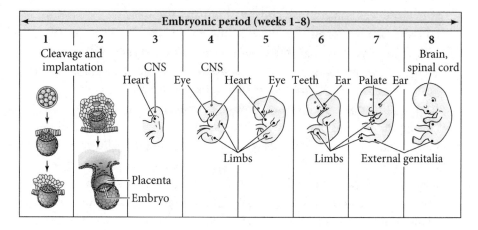

1	2	3	4	5	6	7	8

Embryonic period (weeks 1–8)

Cleavage and implantation — CNS, Heart — CNS, Eye, Heart, Eye — Teeth, Ear, Palate, Ear — Brain, spinal cord

Limbs — Limbs — External genitalia

Placenta
Embryo

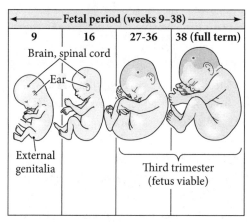

Fetal period (weeks 9–38)

9	16	27–36	38 (full term)

Brain, spinal cord

Ear

External genitalia

Third trimester (fetus viable)

FIGURE 1.1 Human Gestation.

The embryonic period is considered the period of body construction, whereas the fetal period is generally considered that of growth. Some systems, such as the nervous system, keep developing, even after birth. Environmental or genetic damage during the first two weeks of gestation usually affects the entire embryo, and the woman may miscarry without even knowing that she had been pregnant. After the first two weeks, different organs are susceptible to environmental substances at different times.

Source: Gilbert et al. (2005).

Many of the embryos and fetuses seen by the public come from the cameras of Swedish photographer Lennart Nilsson (1965). These images made a sensation when they were first published in 1965 and then reprised in 1990. Like the photograph in *Newsweek*, the fetuses were often taken out of natural context and portrayed as independent entities. One must remember that one cannot take a picture of an uninterpreted human fetus. One would have to either get inside a woman's body or use an aborted fetus. Nilsson used aborted fetuses, coloring and posing them to look like living human fetuses, manipulating the dead hands as well as our emotions (Kevles 1998). As Buklijas and Hopwood (2010) state:

> Although claiming to show the living fetus, Nilsson actually photographed abortus material obtained from women who terminated their pregnancies under the liberal Swedish law. Working with dead embryos allowed Nilsson to experiment with lighting, background and positions, such as placing the thumb into the fetus's mouth. But the origin of the pictures was rarely mentioned, even by "pro-life" activists, who in the 1970s appropriated these icons.

In the *Life* magazine article featuring Nilsson's work, the series of photographs begins with fertilization (illustrated by a picture of in vitro fertilization; Nilsson did not put a camera into a woman's oviduct). The series of photographs ends with a picture of a human fetus (actually, an abortion) with emphasis on its fingers and eyes. It has no umbilical cord, amnion, placenta, or mother. They have all been eliminated. This fetus is a figment of the photographer's imagination and craft. It is not a product of nature.

Pictures are powerful. In 2003, the Supreme Court of the United States upheld an Indiana statute mandating that all women seeking abortions have a one-on-one counseling session during which time they are shown pictures of embryos and fetuses. The abortion lobbyists called this a great victory, since, they claim, such photographs would show the women what they are intending to destroy and convince them not to have the abortion. Demonstrators against abortion routinely hold signs containing pictures, not words. The pictures are the message, and written and verbal argument is irrelevant.

Moreover, many captions describing the pictures are merely spells being cast on you. One widespread photograph on the web shows a photograph of a six-week-old human embryo held in gloved hands. On one anti-abortion website, the embryo is called "a tiny unborn baby." This isn't really true, because this embryo was taken from a tubal pregnancy (which occurs when the embryo adheres to the oviduct instead of to the uterus). This embryo would have killed the mother before it could be born. So this is no "unborn baby." To call it such is a spell, not a description.

The Hogwarts counter-spell to this is nothing less than "Protego." One has to have a counterargument. One has to actually know the science.

READING SCIENCE

We have covered a lot of ground. What we have learned is that biology is conveyed to the public (and to other scientists) by means of metaphors, similes, analogies, and images. This is the way microscopic objects and events can be described so that people can understand them. They make the unfamiliar familiar. It is not surprising, then, that the metaphors, similes, analogies, and images should come from the larger society. So sperm and egg become man and woman, respectively, and their union becomes a marriage. DNA is said to be soul, and the images of human fetuses have been posed, cropped, and described to emphasize such mature attributes as independence and autonomy.

Our reading of science does not deny that sperm or egg exists. But it does recognize that the way sperm and egg are described, the way that DNA and cells are described, and the way that embryonic stem cells and embryos are described depend on language and images. And language and images are culture, not nature. Our perceptions of biological entities are going to depend on what our culture allows us to know about them. This is why multicultural diversity is beneficial to science.

However, as the saying goes, "Everyone is entitled to their own opinion, but not their own facts." So, it is critical to know what the "facts" are (as far as evidence allows) and what has been excluded from being a fact.

This is why we need a "conceptual detox." We have to take a step back and see which stories are "real" and which are spells cast upon us. In this book, we will attempt to provide good science: science with evidence. This does not mean that science is free from cultural influences. No science can ever be free from culture or cultural stories as long as it is done and interpreted by human beings. But while science is "open-minded," it is not "empty-headed." Just because there is no definitive answer does not mean that any alternative is equally good. Some evidence is better than others. As Douglas Adams (2002, p. 98) duly noted, "All opinions are not equal. Some are a very great deal more robust, sophisticated, and well supported in logic and argument than others."

It is often claimed that good education should allow one to recognize excellence in whatever form it might take. What is equally important is that such education should allow you to recognize nonsense and bullshit, no matter how well it is packaged. That's largely what science is about.

2

STORIES OF INFERTILITY
AND ITS CONQUEST

The Sisterhood of Bloody Mary

CLARA PINTO-CORREIA

He who has a why to live can bear almost any how.

—Friedrich Nietzsche, *Twilight of the Idols*

It was early in the morning at my university in Lisbon, July was coming to a close, and the cafeteria was just about empty. One of the young ladies working at the counter saw me finishing my espresso and came over in a hurry, asking for a brief private word. We sat outside for a smoke in the early sun, and she asked me right away what exactly an ICSI[1] was. She was exhausted from having endured so many pointless hormonal cycles before her husband had to face up to the fact that the fertility problem was his, not hers. She would just as well give it all up, but he wouldn't let go. Everybody in their family had children except for them.

OUR FIRST REAL STORY OF SOCIAL PROBLEMS
AND EMOTIONAL COSTS

Her story was not at all unusual. However, it is an important story, as it is an attempt to make sense out of what happens when the story of fertilization does not have a happy ending. Hers is the story of what most hardworking childless young women go through:

The couple had already spent a lot of money on infertility treatments, and borrowed just as much from their parents. The man would never admit that there was something wrong with their fertility, let alone with his sperm. Therefore, they were using a clinic far away, so that no one would know that they were going there. He had to have a child to have a family like everybody else's, but it had to be *his* child. Therefore, simpler and much cheaper solutions, such as adoption or artificial insemination (AI) by an anonymous donor's sperm, were totally out of the question. Now, finally, there apparently being no other way for the couple to have their own precious baby, they were going to get a bank loan to go ahead with ICSI, the injection of a sperm into her egg and the implantation of that developing embryo into her uterus. She was reciting this litany of events as though she were nothing in the whole story but a mere vessel for embryo transfers, traveling by herself the long distance from Lisbon to the clinic in suburban trains, taking hormones and injections, and feeling miserable all the way. Tales of this sort might not be fair, but still they are common, especially among those patients who just hardly manage the money needed for IVF (in vitro fertilization) treatments.

I run an infertility helpline, so I had already heard all this many times before.

Then I got taken by surprise.

I had never heard a young woman caught in these misfortunes tell me that she and her partner were going to leave their present home and move to a smaller one on the other side of the river because her husband couldn't take it anymore: one by one, all the neighbors in their building were having children, and by now they were the only ones left with nothing to show on weekends.

She shrugged her shoulders. "He says that this other building has plenty of old retired people living in it. Besides, and this is important, you know, if we move there, we don't have to pay rent anymore. The apartment belongs to his parents. Maybe we can even just take the boat and the subway to get here; that's what he says. We could sell the car. If we're going to pay back that loan, we can't afford paying rent and gas."

Her husband worked the main gate for the security company and was always extremely nice to me. They were young and had low-income jobs. Like most couples trying infertility treatments in Portugal, they were not being followed by any sort of counselor. With what kind of words could I take upon myself the tremendous responsibility of telling them to stop right there and seek other alternatives? I knew all too well what they were up against. I, too, had endured those infertility treatments, returning each month with new hope, only to be frustrated again. Living with infertility comes with a long, worldwide history of isolation and rejection. And, therefore, how righteous can we be when we tell somebody else to give up their quest and live with what everyone around them perceives as a failure?

ASSISTED HUMAN REPRODUCTION HAS SERIOUS RISKS THAT NO ONE DISCUSSES

I am a developmental biologist, and I love this field. The way I see it, assisted reproductive technologies (ART) came from great basic science and became great biotechnology. Our hard-earned knowledge of how fertilization and early development occur has shown us ways to circumvent nature's blocks to fertility. Whether one uses the comforting metaphor of science "assisting" nature (as in "assisted reproductive technologies") or the more macho metaphor of science "defeating nature's obstacles," ART has allowed thousands of couples to become parents, to deliver healthy babies when nature alone would not. To these couples, ART has been a blessing. But to others, it has been a curse. Like other treatments, it is powerful, and therefore it is dangerous.

Assisted reproductive technologies are a risky tightrope to walk. It can take you from the barren sands of the desert to the milk and honey of the promised land. But, for no particular reason, even if you do everything right, the procedure is usually not successful.[2] You still might be sent back to the desert. Every time this happens, you know right away that your only choice to get that baby is to walk that risky tightrope again. You may choose to do it and live to tell the tale once, twice, or thrice—but you may be getting yourself so addicted to the incredible adrenaline rush of your tightrope walk that you don't even notice that you are becoming both ill and bankrupt in the process. It keeps happening all the time (Couzin-Frankel 2015).

I remember just not wanting to stop because, regardless of all the hassles and all the nausea, the whole ride just felt so exciting, so promising, so *good*. Comparing this experience with other women, many reported without any hesitation *feeling high on pregnancy* during each cycle. There's a baby at the end of this tunnel. And then you want people to stop easily? I stopped when my fourth straight cycle failed because I crashed with a major depression. Of course, it was not pleasant. But I still look back at the experience as a godsend. Before the depression hit, I had been dangerously ready to continue.

When couples enter the clinic, the staff tells them how much the whole procedure costs, and sometimes what the payment options are. But no person is there to tell you just to stop and to consider other ways of going ahead with your lives—before you lose your emotional stability, certainly, and sometimes also before you lose your chances of living your entire life with financial dignity.

SHOULD WE CONSIDER ART AN "INDUSTRY OF GOODS?"

"The spectacle of someone trying to have a child can be more inflammatory than the spectacle of someone trying not to have one," wrote journalist Liza Mundy in 2007 (in Wilson 2014, 17), prefacing the manifestos of those who would later liken the ART industry to that of luxury goods. "Inflammation" is a good biological metaphor of this process. Inflammation occurs when cells rush into a traumatized area in order to cure an affliction, but in so doing, release compounds that cause more damage. "Luxury goods" as a metaphor for human embryos is one rarely used by scientists, but it is an important one, especially when embryos are desired but difficult-to-get items that are easier for the wealthy to acquire. To be considered a "luxury good" puts the human embryo into a framework of a capitalistic economy, not a framework of science. Each framework allows different stories to be told. Can this be possible?

The late twentieth century has witnessed a scientific gold rush of astonishing proportions: the headlong and furious haste to commercialize genetic engineering. . . . The work is uncontrolled. No one supervises it. No federal laws regulate it. There is no coherent government policy, in America or anywhere else in the world. . . . But most disturbing is the fact that no watchdogs are found among the scientists themselves. It is remarkable that nearly every scientist in genetics research is also engaged in the commerce of technology.

This quotation is not from a left-wing article or student activist document. It's from the introduction to Michael Crichton's (1990, ii) *Jurassic Park* and describes the realistic context in which Crichton embeds his modern Frankenstein story. While government support for the integration of physics and commerce became realized earlier in the century as a result of World War II, the promotion of a biomedical industry came in 1980, when the Supreme Court of the United States allowed General Electric to patent genetically modified organisms. The same year, the Bayh–Dole Act in the United States Congress encouraged scientists to make profits from federally funded research. "By 1986," continues Crichton, "at least 362 scientists, including sixty-four in the National Academy, sat on the advisory boards of biotech firms. The number of those who held equity positions or consultancies was several times higher." Biological science had changed.

In other words, yes, there is a definite economic side to having a baby through ART. Assisted reproductive technology firms are profit-making entities whose existence depends on investors, and which seek to maximize profits, not health or happiness. Like any other business, biotechnology companies have an interest in telling stories that make us want to have what they can give. They can provide infertility treatments that allow sterile couples to have babies. For a price.

Whenever a new treatment comes out on the market—even if it is still experimental, even if it is offered without much scientific evidence—those who are infertile flock to it. At some point, long after my early July morning conversation with the young lady from the cafeteria, a new company announced that it could allow women to continually make new egg cells, something scientists doubted was possible. All a woman had to do was to put down $25,000 for their first treatment. Websites about this technique promptly sprang up among those whose hopes had been previously thwarted. One woman posted that she was already $300,000 in debt from her previous treatments, adding, "Yes, that's right, not a typo." Another wrote, "I am betting my life that I will have a baby." One of the company's doctors, who actually told those seeking his services that the miracle-technique in question was not scientifically accepted, was told by many women, "I would do anything to have my own baby" (Couzin-Frankel, 2015).[3]

THE UNIVERSAL CURSE OF INFERTILITY

As we shall discover, the seemingly endless possibilities of ART push young childless couples[4] to dig themselves further and further into one sort of despair or another, whereas women who *choose* not to have children are dismissed as misfits, as traumatized, frigid, careerists or lesbians—some sort of menace to society anyway. And they notice. My support group hot-line in Lisbon was bursting with the frustration of these women. But then again, various forms of despair, together with social ostracism caused by a sense of menace, have always been the fate of childless women, regardless of place, religion, or time. In the United States, infertility among the Puritans was seen as a punishment for religious lapses, to the point where even worldly interventions like herbal remedies were not permitted (Wilson 2014, 20; Marsh and Ronner 1996). What pushes people toward any new sort of medical hope for infertility is something older, more invisible, and certainly more dangerous than just the lure of promising novel technologies for having children.

If we seriously mean to get to the root of the problem, we have to be willing to see what generally is not visible: The women now enduring the grinding treadmills of IVF to no avail are simply repeating the path long traveled by their ancestors who tried their luck with the most unspeakable infertility treatments, all over the planet.[5] And they have always done so, to a large extent because, regardless of how many thousands of years have passed since the first verses of Genesis were written, there is still not one social group, not one culture anywhere on earth, that doesn't abhor infertility. There is a serious reason why the first prayer mentioned in the Bible is Hannah's heartbreaking plea for God to open her womb. Her friends have babies, her husband's other wife has a child; why can't she have one? So fervent is her prayer that the priest who sees her scolds her for being drunk. Just like that priest, those watching the desperate things infertile women do often think we're crazy.

MYTHS AND HISTORIES CONCERNING BARREN WOMEN

The history of civilization and the pantheons of faith alike are full of infertile women, often eventually delivered of predestined children after complex, elaborate plots. All religious, historical, and mythological books have their fair share of goddesses, saints, horseback-riding heroines, and queens and princesses undergoing some momentous transformation because they cannot have children. Uncannily following in the footsteps of primeval worldwide mythologies, history ranging from West to East unfolds as an sequence of infertile empresses, queens, and princesses enduring divorce (think of Princess Soraya of Iran), exile (Mary Tudor and Queen Joana of Portugal), and insanity (Mary Tudor is only the most blood-chilling case) as punishment for their inability to produce children.[6] In Britain, Mary Tudor's desire to have a child (and give England an heir) is fundamental to her mythic history, and her transformation from "Gentle Queen" to "Bloody Mary" is intimately linked to this frustration.

The world's great literature is filled with scenes of desperation over a family chain that is brutally interrupted when the last male heir can no longer reproduce for one reason or another. Indeed, a fresh reading of *Lady Chatterley's Lover* provides some interesting commentary on this tradition. In any century and place, not having children is still a couple's ultimate shame.

We know that European history is full of these tales. As for the history of American colonization, dispensing with queens and princesses didn't really make the picture any brighter. Girls were often married virginal and uninformed, not knowing much about their wedding night other than that they should be ready to endure serious pain. They did feel pain, indeed. However, if this first part of the story sounds too familiar, its particular

American development comes with a blood-chilling note repeated several times in contemporary medical records (Marsh and Ronner 1996). As it happens, it was not all that rare for the pain of these newlyweds to increase as days passed after the nuptials, even though they were not continuing to have intercourse, and, as their condition worsened, their fervently dreamed-of pregnancies never came. When the doctor finally showed up and took a look at a new bride, he realized that her husband had infected her with gonorrhea and that their union would probably be rendered infertile.

These powerful combinations of mythology and factual history can certainly help us understand somewhat better why it is that most modern infertile couples hold on to ART so ardently: Thousands of years later in the course of the history of ideas, traditional childbearing still appears to be the only way of building a family that doesn't reveal to the outside world anything plainly accursed about yourselves, unconscious though these responses to infertility might be in our days.[7]

And, should these couples have counselors, these therapists will often offer the old refrain that passing *your own* genes to *your own* children ultimately represents our age-old desire to live forever by passing our specific traits along to the next generation.

Please, somebody, please press the reset button—if for no other reason than to give us the chance to think differently. We need room to question, to think differently, without our thoughts being framed by this notion of "genetic legacy."

This refrain of the importance of passing on one's genes can become maddening. Since the second half of the last century, it has even become fashionable for sociobiologists and evolutionary psychologists to tell us that our bodies are merely survival machines for our genes, which ache to be perpetuated into the next generation.[8] Television ads bring us messages from companies who will assess our DNA to find out "who we really are" (as if upbringing, wealth, and luck had nothing to do with it). Although this ersatz sociobiological idea has been discredited in science, it is still a popular myth being propagated in supposedly scientific books. Infertility is breaking the great lineage wherein we are connected to the rest of nature. Seemingly, breaking that chain destroys such a powerful yearning that most of those affected are led into strange acts of desperation—as in the story of the young couple at my university, already exhausted, tense, and in debt, but still willing to get a bank loan and to move to a smaller, more distant house to simultaneously save money and save face.

MIRACLE TECHNIQUES AND SOCIAL IMPATIENCE

The importance of saving face should not be trivialized. Infertile women are being shamed for what they cannot do, shamed for failing their womanhood, their husband, their tribe, or their god. And now, it turns out—in the age of technology, we are also failing our science. And everyone around us says that there is definitely no excuse for this.

People hate us for this. I have experienced this hate, and I have lived to tell the tale.

My personal struggle with infertility became serious in the mid-1980s, eventually leading to a number of failed IVF treatments in 1998. People all around me were growing more and more impatient. At first, I assumed it was just my wild imagination telling me this. But it wasn't. There was a generalized attitude of finger-pointing. I was told that "with so many medical techniques available now, there's no excuse you're not pregnant yet, and with triplets at that." But I was teaching embryology in medical school. I was writing and giving lectures about ART. I obtained a Ph.D. in mammalian fertilization. But no baby. I was even doing research in mammalian cloning. Therefore, I ended up believing that people around me were impatient because, with so much knowledge of reproductive biology, I had no excuse to fail so miserably. I tried not to make much of this constant pressure until I started talking with other women in the waiting room of the IVF clinic I attended.[9] And this was how I discovered that all of them were living with the same impatience, the same veiled accusation of not trying hard enough, the same social expectation of medical miracles becoming manifest in the short run, since everybody knew that now such miracles were possible—always.

As the decades unfolded since the first "test-tube babies" of the 1970s, people believed more and more that there was a ready-made remedy for all kinds of affliction that could cause infertility—and that such remedies were always going to work, because medicine always works. If medicine could ward off diabetes and allow people to live into their eighties, could anyone honestly believe that some people are left unable to have children if they try hard enough? Cancer survivors have had to endure endless rounds of gruesome treatments, and it is commonly believed that having a positive attitude and believing in a cure is a defining and crucial part of their survival stories. So why are couples who have endured IVF but come out empty-handed lagging behind? We sometimes hear formerly infertile new mothers exult that God has answered her prayers and that she now has the baby she had prayed for. What is our problem? Are we not religious enough? Not optimistic enough? Not wealthy enough? What?

If those around us rush to judgment and assume that we are not enduring enough treatments or not being positive enough, and therefore that a substantial part of our failure is our own fault, we can't really blame them for this extra weight they dump on us. People outside the field of reproductive biology don't realize that most fertilized human eggs don't survive to become babies. And most people outside the field don't realize how little we actually know about human conception. They hear only the success stories. If the general public knew of both the rarity of a pregnancy coming to term and the vast number of secrets that the physiology of reproduction still holds back from us, it is quite reasonable to argue that society would certainly be much more supportive of unsuccessful IVF patients who often try their luck again and again for an abysmal number of years. It's not the fault of the accusers: Nobody provides them the information they need in a format they can grab onto. For the most, as far as ART goes, people have been given those much juicier myths: counterinformation that makes for great media content but later great social discontent.

Sixth-graders who have studied fertilization in class for the first time can understand the shortcomings of ART in less than one hour when their teachers invite me in for conversations on the subject. Similar to my classroom discussions, this book is trying to provide the information, both the scientific facts and the emotional wisdom, that will make those with children rejoice in their good fortune and, at the same time, give couples without children empathy, patience, and understanding. Without such understanding, we would have to agree with Aldous Huxley's famous aphorism (1937, 8): "Technological progress has merely provided us with more efficient means for going backwards." If, owing to technological progress, society is now passing severe judgment on those who remain childless because they certainly could try harder[10] then vast numbers of people with already broken hearts are suffering even more than they were before the 1980s, when such reproductive technologies first became available. Such backsliding could easily be prevented. But, so far, it has not been.

MEDICAL COSTS AND BIOLOGICAL DIVIDES

The story of the young couple working at my university brings to mind yet another issue that offers no moral comfort at all: an alarming reality that will also have to be called into question among these pages.

These two young people were both from families with limited means. Still, they had some funds that they could spend in their quest for children, and their parents were willing to lend them just about all the money they could to back them up—likely not expecting ever to be paid back, as so often happens with parents' generosity. And, when paying for ICSI was more than all these efforts combined could afford, the couple might have been perceived as sealing a pact with the devil when they somehow qualified for a bank loan. There is a lot to say about the price people are willing to pay to have children.

As we look at social and emotional impacts, we will necessarily explore some of the many situations women face in trying to have a baby who won't come naturally. It doesn't take long to notice that a lot of "zoning" goes on in ART. There are those who have to get in line and wait their turn for the publicly funded services offered by their countries. And that waiting often takes years owing to a lack of clinical resources, which is not ideal, since most women only resort to ART in their late 30s and therefore don't have all the time in the world.[11] There are those who do not have any free public services available nearby, whereas extravagantly posh private clinics abound—for the foreigners who can afford them. There are those who live light years away from infertility clinics, although they know they exist, as do their neighbors. There are those who live in countries where entering an infertility clinic is akin to risking one's life, owing to the risk of dangerous infections or unscrupulously made drugs used for treatments.

Still, if the only alternative is to remain childless, many women will take that risk any day. There are even those who make a very good, decent, or frankly meager living by selling their fertility, their wombs, or both, to infertile patients from around the world who are rich enough or desperate enough to pay for these services.

And there are also northern European countries, where population decline and functioning welfare states have combined to offer citizens all sorts of free ART treatments, often for more than one cycle and without any exception. These are some of the countries where the entire population enjoys the highest living standards in the world.

This picture can't be seen as anything but troubling, because biomedical fences between rich and poor are increasingly becoming a hallmark of this century. If we are to seek a ready-made example of an area where the gap between the haves and the have-nots keeps widening as scientific progress allows for more and more options, ART provides a simple and upsetting example. In countries such as the United States, however, this divide can still be considered just a small gap. On the other hand, thinking instead of the entire planet, and of what the vast majority of infertile women go through in stoic silence, from China to Egypt to India and stretching through sub-Saharan Africa, we see that this gap promptly becomes something more like a rift with the poor going left and the rich going right with a huge biological void in between, slowly growing in size from one day to the next, with no agency to regulate what to do or any consensus even on what *should* be done.

TEST TUBES IN MY BOSOM: WHY I NEED TO WRITE THIS BOOK

In the previous chapter, Scott talked about why *he* had to write this book, and how we are being taught the wrong stories of fertilization and embryonic development. Now I have to explain why *I* am writing this book. More than anything else, I'm on a quest to destroy those myths about infertility that arise from ignorance and prejudice. This particular battle has been made unnecessarily difficult by the ways that those who govern have poisoned our minds by hindering access to knowledge and tainting our interactions by making us objects of competition. My quest started when I noticed that adults were asking me questions that indicated they had no idea what goes on either inside the body or during assisted reproduction. I began this chapter with a personal story; let me end it with another one. It features a cultured, well-educated, intelligent person I truly love and admire. But, like the relatively uneducated couple I had spoken with about ICSI, he was clueless about human development and ART. Even the educated elite are imprisoned in near total ignorance about human development and assisted reproduction.

When I was undergoing my infertility treatments, a friend of mine, a successful architect from Australia visited me and asked, "So, where do you keep your test-tube babies while you wait for them to hatch? I always thought you would have them in full view, in some kind of fancy altar installation with lots of candles, right here in the living room."

This man was a longtime companion I knew well, and I knew he was intelligent. So I logically assumed he was joking, and so I joked back: "I always carry them in my bosom, you know. I have to rotate them gently every forty-five minutes in order to maintain the level of both heat and humidity surrounding the test tubes constant, so that the babies hatch in perfect health. Ever since the doctor gave me my test tubes already filled with my tiny little kids, I've been wearing a Wonderbra."

Much to my surprise, he didn't laugh at all. He just asked, in a distracted voice, "How come you never break any test tubes when you fall asleep?"

Ha. He was playing it smart.

"Well, back in Roman times, Livia Augusta, one of Nero's wives, having read Pliny, wanted to see if she could hatch a chick egg in her boobs. When she went to bed, she turned the task over to one of the slaves she trusted most, as long as she had a good bosom for the job. Since I have no slaves, I asked my girlfriends for help. They are all more than willing to oblige because this can be great fun. Then on Sundays there is a contest for the best incubating bosom; our partners act as jurors, my husband is their president, and the winner gets to go to the casino and put down all she wants at the roulette wheel. We all get to go and cheer, including the test-tube babies, because it is good to expose their forming brains to as many stimuli as possible right from the start. Ah, and believe me, we do dress to kill."

For the first time, he finally gave signs of something resembling disbelief.

"Is that really a true story?"

He had turned around to look back at me. I could tell from his face how hopelessly perplexed he was. It was my time to freeze. This highly sophisticated grownup, capable of designing buildings and directing construction sites from Adelaide to Helsinki, had at first not doubted a word I said. The plot had just become a bit too wild with the jury of partners for the best incubating bosom, and then finally derailed with the test-tube babies inside Mom's Wonderbra watching her all dressed up at the roulette wheel. What if I had not gone that far and been content with the Livia Augusta example? We are an ignorant society, and those people who have not experienced ART are often totally unaware of what these technologies entail.

The time for this ignorance must end. We must learn how to tell our stories to our friends, our children, and ourselves. We must give people some real information about fertilization, human development, and ART, and start real discussions of the many critical issues infertility raises. I hope that the chapters in this book, based both on sound scholarship and on profound human stories, will persuade you of the urgent need for education, regulation, and even some international consensus in the domain of ART. Given the global and fragile patchwork nature of the way we all live together on this planet right now, we do not enjoy the luxury of leaving debates of such magnitude forever unsettled.

II

FERTILIZATION AND ITS DISCONTENTS

—

I n this section, we study misconceptions at two levels. Chapter 3 looks at misconceptions about how we depict fertilization. The story of fertilization is often told as a man's story, a competitive race ending in the capturing and penetrating of an egg. Science finds these stories to be false myths. The human female reproductive tract is not a raceway, but an active set of organs that give sperm the ability to activate the egg; and the merging of sperm and egg is done not by boring or drilling through the egg, but by the melting of cell membranes so that the two cells become one. Thus, the data actually suggest a complex series of interactions among sperm, egg, and female reproductive tract, where both egg and sperm are at times active and at other times passive.

The second set of misconceptions involves the physical barriers to conception. Artificial insemination (AI) and in vitro fertilization were invented to circumvent these blocks. From royal families to the general public and even to farm animals, AI has had many guises. One of these is sperm banking, in which women pay to be inseminated with sperm from men having the traits they desire. But it does not always happen that one gets what one pays for. In vitro fertilization offers hope, the source of both blessings and curses, to many infertile couples. The blessings are successful fertilization, pregnancy, and birth. The curses are less talked about: depression, divorce, and bankruptcy.

3

FERTILIZATION

───

Two Cells at the Verge of Death Cooperate to Form a New Body That Lasts Decades

SCOTT GILBERT

When any real progress is made, we unlearn and learn anew what we thought we knew before.

—Henry David Thoreau

THE IMPORTANCE OF UNLEARNING

Where did you learn about the birds and the bees? Who taught you about fertilization? What images do you see when you fantasize about the sperm and the egg? Take a moment to think about how you visualize fertilization, and try to remember why you think this is the case. Most of what you learned about human fertilization and early development may well have come from parents, classmates, teachers, filmmakers, priests, ministers, rabbis, imams, cartoonists, and journalists, many of them having no knowledge of what they were talking about. Many were consciously integrating the sperm and egg into existing dramas, casting spells on us to make us think that fertilization is a competitive adventure among sperm, that the egg is a passive prize awarded to the victor of this competition, and that fertilization takes place immediately after intercourse.

And here is one of the first places where we have to unlearn a great deal of falsehood, for each of those statements about fertilization is wrong.

Each is a spell cast on us to make us think about nature and our bodies in competitive ways. What scientists know is that human fertilization is a process involving amazing feats of cooperation among cells, men, and women:

- First, the race is not always won by the swiftest. The sperm that get to the egg first are not usually the ones that fertilize it. They are still immature.
- Second, the female reproductive tract is not passive. Rather, the cells of the oviducts ("Fallopian tubes" or simply "tubes") bind to the sperm, slow it down, and secrete proteins that change the cell membranes of the sperm. These cell membrane changes allow the sperm to fuse with the egg to generate the one-celled embryo (the zygote). Thus, the sperm collected in condoms are immature, incapable of fertilizing eggs, because they have not been modified by the cells of the oviduct. In this regard, they are like any other body cell.
- Third, the sperm does not bore or drill its way into the egg. Indeed, the sperm "spoons" with the egg, lying next to it. At the point of contact, the cell membranes of the egg and sperm melt together, causing the two nuclei to be in the same cell. The sperm and egg cells are both active in this process.
- Fourth, fertilization takes time. Fertilization does not occur in a moment of passion. It occurs four or five days after intercourse, when the woman is reading, watching television, or working. Intercourse is not "impregnation."

So let's look more closely at the events of fertilization, starting with the two major partners in this dance, the sperm and egg (figure 3.1).

THE SPERM

Sperm were discovered in the 1670s, but fertilization was not discovered until the 1870s. So for two hundred years, people didn't know what the sperm did. That, of course, did not stop anyone from speculating about sperm (Pinto-Correia 1997). The small microscopes of Dutch cloth

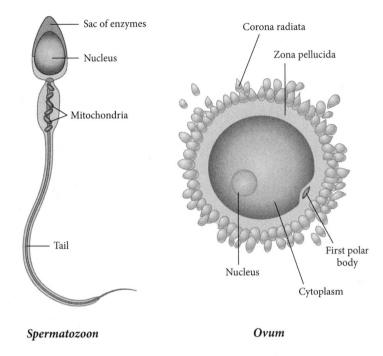

Sac of enzymes

Nucleus

Mitochondria

Tail

Corona radiata

Zona pellucida

First polar body

Nucleus

Cytoplasm

Spermatozoon   *Ovum*

FIGURE 3.1 **Structure of the Human Sperm (*left*) and Egg (*right*).**

The sperm is a streamlined cell. The sperm head contains the nucleus (full of DNA, the hereditary material) and a sac of digestive enzymes that will enable it to reach the egg. It has gotten rid of most of its cytoplasm. The mitochondria provide energy to move the tail for locomotion. The egg contains a nucleus in a large amount of cytoplasm. A polar body derived from the first meiotic division is seen, and it will soon dissolve. The cell membrane is thick, as it contains thousands of enzyme sacs beneath it, and it is surrounded by a zona pellucida, which will allow sperm to bind and which will prevent the egg and early embryo from adhering to the oviduct. A layer of follicle cells also surrounds the egg as it lies in the oviduct. This is not to scale, as the human egg is about 40 times bigger than the sperm.

merchant Antonie van Leeuwenhoek opened up a new world. Leeuwenhoek discovered bacteria, the lice that infested human skin, and, strangely, sperm. He was very careful about this last finding, writing to the Royal Society of London that he did not obtain his samples by sinful acts (i.e., masturbation), but by "normal marital overplus." (You can almost imagine Tony's wife telling him to get back into bed, while the master ran to his apparatus.) In his semen, Leeuwenhoek found a frenzied world of living, swimming beings, undetectable to the unaided eye, which he called "spermatozoa."

Spermatozoa means "seed animals." Leeuwenhoek chose this term because he believed that sperm were like seeds. In other words, they carried in their round heads the preformed infant, what is sometimes called a "homunculus." In this view, the sperm is placed into the woman's body, just as a seed is placed into moist soil, and the penis acts as a trowel. The mother does not provide any hereditary material but merely provides the nutrients and supportive conditions needed to allow the growth of the little infant already present in each sperm. Of course, just as the quality of soil can influence the growth of a seed, so the quality of a woman's uterus could influence the growth of a homunculus. But it was the man's sperm that was the sole bearer of inherited traits. The agricultural analogy of a man sowing his seeds into a woman's soil became the major story of heredity and development. The metaphor of the preformed seed inside the sperm has a long history. Indeed, in 1931, embryologist and historian Joseph Needham pointed out that this idea enabled the conduct of men during war: men were killed and women raped. The offspring of the women were not considered those of the conquered, but those of the conqueror.

But "the little man in the sperm" idea had some problems. One was that if the sperm were little humans and had souls, then each ejaculation caused more deaths than all human warfare combined. Heaven would be populated with the souls destroyed during masturbation and through intercourse. So another hypothesis held that it was the egg, not the sperm, that held all the attributes of the offspring. In this theory, the semen (and the sperm within it) were merely agents that activated the egg. In the 1700s, Lazzaro Spallanzani (a priest who pioneered in vitro fertilization in frogs and dogs) put silk jocks on male frogs so that he could filter the sperm out of their semen. Such filtered semen, he found, was not able to cause the development of a frog egg (Pinto-Correia 1997). One would think that Spallanzani would have declared that he had found that the sperm *and* egg needed to come together. But he did not. Spallanzani believed that the egg, not the sperm, contained the preformed organism. He thought the sperm were parasitic worms and that it was the energy of the semen that stimulated the egg to develop.

It wasn't until the nineteenth century that cell theory and better microscopy showed that sperm were not parasites but were derived from normal-looking cells embedded in the testes. The sperm, it seemed, were just cells, but very special ones. We now know that human sperm derives from stem cells that migrate to and lodge inside the testes. As stem cells, they make more of themselves, and also make a type of cell that becomes the sperm.[1] As the sperm matures, it loses half its chromosomes in a process called meiosis (box 3.1). That is, instead of dividing such that each cell keeps the same number of chromosomes, each sperm cell has only half the normal chromosome number. The developing sperm also gets rid of most of its cell body (the cytoplasm), reducing itself to essentials— a "head" containing the nucleus and its chromosomes, a tail that provides locomotion, and mitochondria, which provide energy to the tail. The sperm also has a bag of digestive proteins at the top of its nucleus, and these proteins help the sperm reach the egg late in their journey. Sperm cells are being made all the time, every second of the day. From the onset of puberty, men make millions of sperm each minute.

BOX 3.1: MEIOSIS AND SEX DETERMINATION

Sexual reproduction is nature's masterpiece. It is the basis of biodiversity, variation, and the continuity of life. Sexual reproduction combines two of nature's most powerful forces: sex and reproduction. *Sex* means the recombining of genes. In sex, the sperm and egg each bring their halves of the genome to a new individual. You are not all from your father or all from your mother (as would be the case in cloning). Rather, you are a 50:50 mix of nuclear genes, half from mom, half from dad. *Reproduction* means the making of new organisms from older ones. Putting these processes together in sexual reproduction means that newly produced organisms are different from their parents. This creates new variations in each generation. It is the basis of biodiversity and evolution.

The mechanism by which the sperm and egg reduce their chromosome number by half is called meiosis (box figure 3.1A). In humans, each

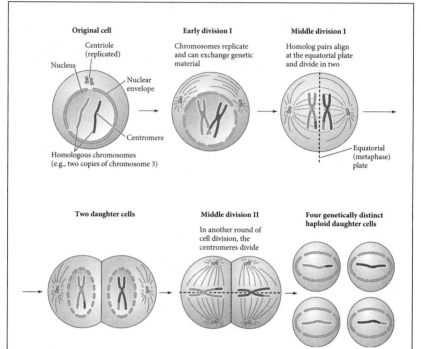

Original cell

Centriole
(replicated)

Nucleus

Nuclear
envelope

Centromere

Homologous chromosomes
(e.g., two copies of chromosome 3)

Early division I

Chromosomes replicate
and can exchange genetic
material

Middle division I

Homolog pairs align
at the equatorial plate
and divide in two

Equatorial
(metaphase)
plate

Two daughter cells

Middle division II

In another round of
cell division, the
centromeres divide

**Four genetically distinct
haploid daughter cells**

BOX FIGURE 3.1A Meiosis.

During the first meiotic division, homologous chromosomes from your father and mother pair up. (For simplicity, we represent only one chromosome in the figure.) Chromosome 1 from your mother pairs with chromosome 1 from your father. They replicate their genes, making a total of four sets, two from your father's chromosomes, two from your mother's. In the first division, these sets get separated. Some cells get your father's chromosome 1; some cells get your mother's chromosome 1. Each cell has different sets of chromosomes from your mother and father. The second meiotic division separates those pairs into single chromosomes. Thus, each sperm and egg cell has half the number of your chromosomes (the "haploid" genome), and you transmit a different assortment of chromosomes from your mother and your father. At fertilization, the half-genome from the sperm (23 chromosomes) meets the half-genome from the egg (23 chromosomes) to make the normal 46-chromosome human genome.

Source: Gilbert et al. (2005).

nucleus contains forty-six chromosomes: twenty-two pairs of regular chromosomes, and two chromosomes that determine sex (called the X and Y chromosomes). Meiosis consists of two cell divisions. During the first meiotic division, the chromosome pairs come together. In other words, the chromosome 1 that you receive from your mother pairs with the chromosome 1 you receive from your father. Similarly, chromosome 21 from your mother and chromosome 21 from your father find each other and lie next to each other. Each chromosome then makes a copy of itself. In other words, the DNA that is the basis of each chromosome makes a second chromosome out of material from the cell. So there are now four small chromosome 1's, as well as four copies of every other chromosome. Two of each foursome are from the mother (and are bound together), and the two others are from your father (and they, too, are bound together). The first division of meiosis separates these pairs randomly. Thus, a daughter cell might have the chromosome pair of your father's chromosome 1 and the chromosome pair from your mother's chromosome 21. Since there are 23 pairs, there is only a remarkably small chance that a cell would receive all its chromosomes from your mother (your child's grandmother) or all its chromosomes from your father (your child's grandfather).

The next cell division, the second meiotic division, breaks the bonds separating the pairs. The result is a set of four cells, each containing half the normal number of chromosomes (twenty-two normal chromosomes and one sex chromosome) and each containing a different set of chromosomes. At fertilization, the sperm with its twenty-three chromosomes meets the egg with its twenty-three chromosomes, and the normal number of forty-six chromosomes is re-established. But the forty-six chromosomes are different (i.e., have different genes) from either of the parents.

Meiosis differs in males and females (box figure 3.1B). Meiosis in males generates four equally sized cells that will each become sperm. In females, instead of dividing equally, one cell receives nearly all the cytoplasm, while the other three become "polar bodies." A polar body is just a nucleus encased in a thin band of cytoplasm. At each division of the egg precursor cell, the cell volume is conserved. Thus, the egg cell maintains its big size, whereas the sperm grows smaller.

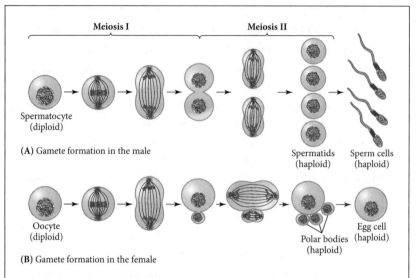

Meiosis I   **Meiosis II**

Spermatocyte
(diploid)

(A) Gamete formation in the male

Spermatids   Sperm cells
(haploid)   (haploid)

Oocyte
(diploid)

Egg cell
(haploid)

Polar bodies
(haploid)

(B) Gamete formation in the female

BOX FIGURE 3.1B Meiosis in Males and Females.

In males, meiosis leads to four sperm cells, each with half the normal number of chromosomes. In humans, each sperm has twenty-three chromosomes (twenty-two normal chromosomes and either the X or the Y). In females, meiosis leads to the formation of one big cell, the egg (ovum), plus small polar bodies that degenerate.

The X and the Y chromosomes are critical in determining the sex of the baby. Males usually start off with one X and one Y chromosome. The X chromosome is necessary for cells to survive. The Y chromosome has a gene (called *SRY*) that starts the reactions that turn the embryonic gonads into testes. Females start off with two X chromosomes. Together, the two X chromosomes activate the reactions that turn the gonads into ovaries. Thus, males are usually "XY" and females are usually "XX." Each egg contains an X chromosome (since the female cells are XX). The sperm, however, can be either an X-bearing sperm or a Y-bearing sperm (box figure 3.1C). When an X-bearing sperm meets an X-bearing egg, the child will be XX and usually a girl. When a Y-bearing sperm meets the X-bearing egg, the child will be XY and usually a boy.[2]

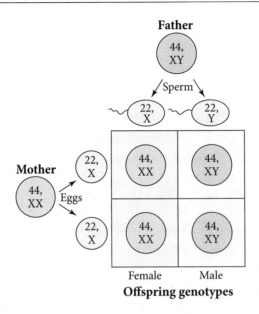

BOX FIGURE 3.1C Sex Determination in Humans.

During meiosis, the eggs produced by the mother each receive a single X chromosome. This is because women have two X chromosomes. The sperm produced by the father can have either an X or a Y chromosome. Thus, whether a baby will start developing as male or female depends on whether the X-bearing egg was fertilized with an X-bearing sperm or a Y-bearing sperm. Those eggs with two X chromosomes usually develop ovaries, whereas those with an X and a Y chromosome usually develop testes. This process leads to a sex ratio that is approximately even.

Source: Gilbert et al. (2005).

So a man produces two types of sperm, and a woman produces one type of egg. There is generally a 50 percent chance that the child will be a boy and a 50 percent chance of the child being a girl. Even if a couple has five boys in a row, the chance of the next child being a girl is still 50 percent.

THE EGG

The egg was more difficult to find. Finally, in 1828, Karl Ernst von Baer, a young biologist working in Estonia, found what other scientists had sought in vain—the human egg. Just as the sperm are made in the testes, the eggs are made in the ovaries. Each woman is born with millions of eggs, but most eggs perish before they have a chance of meeting sperm. This is because the precursors of the eggs in the ovaries are not stem cells and so cannot repopulate or continue to make more eggs. Rather, the egg precursor cells in the ovary divide a few times and then begin their meiotic cell divisions. (The sperm wait until puberty to start meiosis.) Most of the developing eggs have died before a female infant is born. In human females, one or two eggs are pushed out of the ovary every month. This eruption of an egg from the ovary is called ovulation. Of the millions of egg cells formed in the embryonic ovaries, only about five hundred will erupt from the ovary into the oviducts during a woman's lifetime, about one each month after puberty. It is in the Fallopian tubes that the eggs have the chance of meeting a sperm.

But unlike the sperm, the egg, during its development, has blossomed. Its cytoplasm has been enlarged, not reduced. It not only has a nucleus with half the number of original chromosomes, it has a rounded cytoplasm thousands of times bigger than the sperm. This cytoplasm houses the proteins needed for the embryo to grow. It also has the mitochondria to produce the energy needed for cell division. But while the sperm use their mitochondria to furiously whip their tails to get to the egg, the egg mitochondria are waiting to provide energy to the embryo. Indeed, they provide the energy for the adult as long as it lives. All of a body's mitochondria, the part of the cell that uses oxygen to produce energy—come from the mother's egg. None of them come from the sperm.

The sperm and the egg are very different—but they are also very much the same. Unlike any other cell in the body, they have only half the number of chromosomes (where the genes are located). Only when they fuse together will the normal number of genes be re-established. And, unlike any other cell in the body, the cell membranes of the sperm and egg can fuse with each other. And both cells, the sperm and the egg, are on the verge of death. If they don't find each other soon after release from the testes and ovary, they die.

THE JOURNEY OF THE EGG: OVULATION

The developing egg cells (oocytes) mature in the ovary. After puberty begins, one immature egg (and sometimes two) makes a leap into the unknown once each month. In other words, the human egg starts its meiotic divisions in the fetal ovary but doesn't resume this cell division until between twelve and fifty years later, when it is released from the ovary! It won't finish these meiotic divisions unless it gets signals from the sperm during fertilization.

Ovulation is regulated by the hormones of the menstrual cycle. It is important to understand how these hormones work, because our knowledge of the menstrual cycle is the scientific foundation for two critical and opposing reproductive technologies: birth control and in vitro fertilization (IVF). *Disrupting* these hormones is the basis of chemical contraception. Birth control pills work by preventing these hormones from functioning, thereby stopping the menstrual cycle. Thus, there is no ovulation, and without an egg in the oviduct, fertilization cannot occur, no matter how many sperm find their way there. Conversely, *activating* the menstrual cycle hormones to higher-than-normal levels is the basis of the ovarian hyperstimulation protocols used to obtain several eggs simultaneously for IVF. Here, by changing the levels of the menstrual cycle hormones, numerous eggs (often a dozen or more) can be stimulated to ovulate at the same time, rather than the usual one or two. This protocol allows fertility specialists to obtain the maturing eggs needed for IVF.[3]

To understand the menstrual cycle, though, one needs first to understand a bit about female human reproductive anatomy (figure 3.2). The female reproductive tract carries out many functions. First, it provides places for eggs to mature. These are the ovaries, the gonads of females. Second, the female reproductive tract provides a structure wherein the mature egg is fertilized. These are the oviducts. Whereas the testes contain a duct system to take the sperm out of the gonad, the egg erupts from

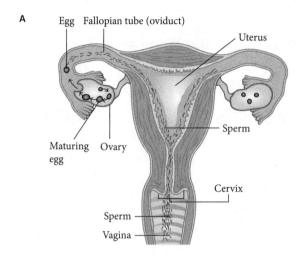

A

Egg Fallopian tube (oviduct)

Uterus

Sperm

Maturing egg Ovary

Cervix

Sperm

Vagina

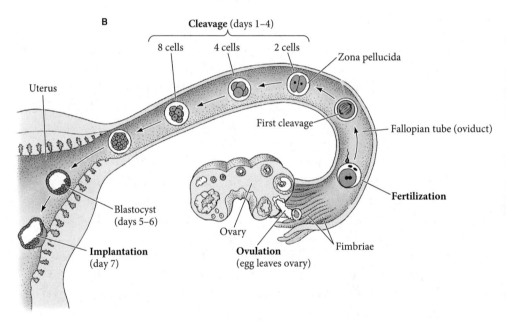

B

Cleavage (days 1–4)

8 cells 4 cells 2 cells

Zona pellucida

Uterus

First cleavage

Fallopian tube (oviduct)

Fertilization

Blastocyst (days 5–6)

Ovary

Implantation (day 7)

Ovulation (egg leaves ovary)

Fimbriae

FIGURE 3.2 **Female Reproductive Anatomy and Path of Sperm.**

(A) Overview of internal female reproductive anatomy, showing the relationships of the ovary to the fallopian tube, uterus, cervix, and vagina. The egg has been ovulated, and sperm are migrating throughout the ducts. Fertilization occurs in the ducts, not in the uterus (as is often believed). (B) Once fertilization occurs, the egg starts dividing and migrating to the uterus, where it attaches and implants. Days are approximate.

Source: Gilbert et al. (2005).

the *outside* surface of the ovary and does not go directly into an oviduct. Rather, it (and some of its surrounding cells from the ovary) is swept into the oviduct, where it will secrete factors that attract the sperm. Third, the female reproductive tract has a place to support the developing embryo. This is the uterus, sometimes called the womb. The oviducts lead into the womb, and the fertilized egg travels down the oviduct, dividing into embryonic cells in a three-day journey. Fourth, the female reproductive tract provides a means for the sperm to travel to the egg. This is accomplished by the vagina (birth canal) and cervix and also by the uterus and oviducts.

In natural human reproduction, the sperm are ejaculated into the vagina. The cervix, at the base of the uterus, can regulate sperm entry into the uterus by the amount and stickiness of its mucus. (Indeed, the amount and stickiness of the mucus are what are being measured in many of the fertility tests and ovulation predictors sold in drug stores.) If the mucus is supportive and made of thin strands, the sperm can enter into the uterus. The sperm then swim into the oviducts. Here, the sperm are activated by chemicals from the oviduct cells, allowing them to fertilize the egg (Austin 1952). This is called "capacitation"—the gaining of capacity—and it occurs just before the place in the oviduct where fertilization occurs. When the egg is fertilized, in a region of the oviduct close to the ovary, it starts dividing and is wafted to the uterus. There, the egg breaks out of the protein shell that enclosed it and attaches itself to the uterus. It burrows into the uterus, attracts blood vessels to itself, and begins to grow.

In addition, the female reproductive system includes external elements. The labia are the fleshy folds that enclose and protect the vagina. They form from the same embryonic region that forms the scrotum (testes sack) in males. The clitoris forms where the folds meet, and it comes from the same embryonic region that forms the penis in men. Unlike the penis, it is not used for taking urine or reproductive cells outside the body. Like the penis, though, it has numerous nerve endings, is sensitive to stimulation, and can become erect when stimulated. The labia contain lubricating glands, and these glands, along with the clitoris, probably function more for pleasure than reproduction per se.

THE MENSTRUAL CYCLE

The menstrual cycle is a set of monthly[4] hormonal changes that integrate the functions of the female reproductive system (Fritz and Speroff 2010) (figure 3.3). These changes coordinate (1) the development of the oocyte in the ovary; (2) the growth of the uterine lining that enables the uterus to catch and support an early embryo; and (3) the amount and stickiness of the mucus in the cervix that regulate whether sperm can enter into the deeper reaches of the female reproductive tract.

The menstrual cycle is said to begin each time a woman begins her period; that is, when blood can be seen coming from the vagina. This blood is the result of the body's shedding the uterine tissue that could have caught an embryo had pregnancy occurred. If pregnancy does not occur, this tissue, and its blood supply, are shed from the body. The pituitary gland at the base of the brain is then instructed to produce increasing amounts of a hormone called follicle-stimulating hormone (FSH). Hormones are chemicals that spread through the bloodstream and can bind to cells in numerous organs. Therefore, hormones are excellent signals to use to coordinate changes throughout the body. The FSH from the pituitary gland stimulates the ovarian follicle cells (which surround each egg) to secrete the hormone estrogen. Estrogen does many things in many organs.

First, the doses of estrogen made in the ovary allow the development of a single egg. The maturing egg and its surrounding cells are called the follicle. (Several follicles may start to develop, but usually only one, "the dominant follicle," finishes. Sometimes, two eggs re-initiate development, and both become ovulated. Both these eggs can be fertilized, and this is one of the causes of nonidentical twins.)

Second, the estrogen made in the ovary goes throughout the blood and instructs the lining of the uterus to grow. These new cells form the endometrium of the uterus, the inside cellular cushion that will catch the new embryo as it is propelled down the oviducts and into the uterus.

Third, estrogen instructs the cells of the cervix to produce a type of mucus that will facilitate sperm entry into the uterus and help the sperm reach the egg.

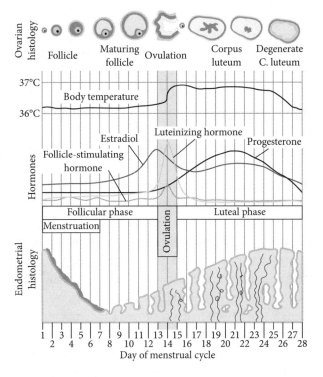

FIGURE 3.3 The Menstrual Cycle.

The first part of the menstrual cycle is characterized by an increase in the level of estrogen, which causes the proliferation of the uterine lining (the endometrium). This is called the follicular phase, and it is also associated with the thinning of the cervical mucus. At high levels, estrogen stimulates the production of luteinizing hormone (LH), which initiates ovulation. The uterus is now prepared to receive an early embryo. If pregnancy occurs, the follicle (now called the "corpus leuteum") produces progesterone, which stimulates the uterus to send blood vessels to the embryo and to allow the embryo to enter into it (the luteal phase). If fertilization does not occur, the endometrial lining of the uterus (and its blood) is released and the "period" begins. Fluctuations in estrogen levels and lack of estrogen also cause body temperate changes, which can predict ovulation and also cause the "hot flashes" of menopause.

And fourth, the ovary's estrogen prevents the release of the developing egg before it is mature. It does this by blocking the pituitary gland's secretion of another hormone called luteinizing hormone (LH). When the egg is mature, the levels of estrogen reach a critical concentration, and instead of inhibiting LH secretion from the pituitary gland, the high levels of estrogen *promote* LH production and secretion. Luteinizing hormone is the hormone that promotes ovulation. It tells the follicle cells to dissolve the protein wall around the egg and forces it to be released into the oviduct. The mature egg, together with some of its follicle cells, is swept into the oviduct by tiny protrusions of the oviduct. The egg now waits for the sperm. It can survive about a day if it is not fertilized. If no sperm are present, the egg degenerates. If sperm are present, fertilization occurs, and small currents in the oviduct push the newly fertilized egg into the uterus.

The LH and FSH cause the remaining cells of the dominant follicle to produce another hormone, progesterone. Progesterone also does several things. One of its major functions is to instruct the uterine cells to change. Whereas estrogen tells the uterine cells to divide and make the endometrial lining, progesterone tells the endometrial lining to become spongy and receptive to an incoming embryo. It also increases the blood supply to the uterus and prevents its muscles from contracting. Meanwhile, progesterone stimulates the uterus to make proteins that may attract the embryo and assist in its docking to the endometrial lining. There is a great deal of cross-talk between the embryo and the uterus, and if this is impeded, infertility could result (Fritz et al. 2014). Another effect of progesterone is to thicken the cervical mucus, preventing sperm entry into the female reproductive tract, thus preventing further pregnancies.

If implantation of the embryo into the uterus occurs, the new embryo will make a hormone called human chorionic gonadotropin (hCG),[5] which promotes the continued production of progesterone by ovary. The uterine lining will be maintained, and the pregnancy will continue. Moreover, progesterone will prevent FSH and LH production, preventing the maturation of other eggs during the pregnancy.

However, if fertilization and implantation do not occur, there is no hCG. The ovarian follicle stops making progesterone, and as a result, the uterine lining is shed. Menstruation begins. And without progesterone, FSH can be made and secreted, starting the maturation of another egg.

Chemical contraceptives function by interfering with this cycle. Most chemical contraceptives are artificial progesterones. As we just mentioned, progesterones inhibit ovulation and cause the cervical mucus to thicken. Depending on the dose of progesterone, contraception may

---

BOX 3.2: WHEN IN DOUBT, HAVE A PLAN B

Morning-after pills (such as Plan B) are emergency contraceptive pills that can prevent pregnancy after sexual intercourse by inhibiting ovulation. They have relatively high doses of a particular progesterone-like compound. Some religious groups of pharmacists do not want to sell this drug, claiming that Plan B is possibly an abortion-inducing agent that prevents the adhesion of the embryo to the uterine lining. Radio commentators such as Rush Limbaugh have called Plan B an abortion pill. As of 2017, however, there are no data supporting this claim.[6] Rather, when detailed measurements of ovulation were taken, it was found that women who had taken Plan B *after* ovulation became pregnant at normal rates (Noé et al. 2011; Vargas et al. 2012). None of the women who took Plan B *prior* to the day they were expected to ovulate became pregnant. Moreover, the progesterone-like compound in Plan B did not cause significant changes in uterine gene expression. Two conclusions were drawn from these studies: (1) The only known mechanism of Plan B is to prevent ovulation, thereby working solely as a *contraceptive* (not producing abortions of existing embryos); and (2) the morning-after pill is not good birth control because if ovulation has already occurred (and most women do not know when this has happened), the morning-after pill will not work. The egg is already in the tube and ready to meet sperm.

---

work by one or both of these mechanisms. In most contraceptive pills, a synthetic estrogen is also added at levels that block the production of FSH, further inhibiting ovulation. (See box 3.2 for one particular form of contraception.)

THE JOURNEY OF THE SPERM

The sperm's journey begins even before it is ejaculated (figure 3.4). The sperm are made in the testes. While follicle cells surround each immature egg cell in the ovary, the sperm mature in small tubes called seminiferous

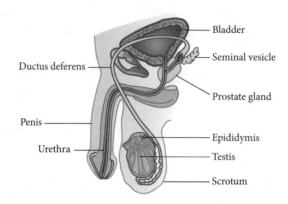

Ductus deferens

Penis

Urethra

Bladder

Seminal vesicle

Prostate gland

Epididymis

Testis

Scrotum

FIGURE 3.4 Male Reproductive Anatomy.

Sperm are made in the seminiferous tubules of the testis and are transported out by a tubule system including the epididymis and ductis deferens. The seminal vesicle and prostate gland provide much of the fluid for the semen as well as chemicals that help lengthen the lifespan of the sperm cells. The spongy cells of the penis expand with blood during erections. The ductus deferens (taking the sperm) opens into the urethra (taking urine from the bladder), so the penis has but one opening.

tubules (a fancy way of saying "sperm-bearing small tubes.") The testes (testicles) are housed in a sac called the scrotum. The scrotum acts as a climate-control system. Special muscles raise and lower the scrotum to keep the developing sperm cells just below normal body temperature. The sperm go through tubes taking them out of the testes into the penis. The penis has many functions. It gets rid of urine from the bladder, and it expels sperm using the same tube. It also functions as an organ of pleasure, having a high density of touch-sensitive nerve endings at its tip. Indeed, the penis is an organ that can simultaneously give and receive pleasure. As the sperm enter the penis, the penis lengthens and hardens owing to blood flowing into its tissues. Neural stimulation (either physical or mental) enables blood to flow out of the penis capillaries and into the surrounding tissues, greatly expanding them. This is critical for the penis's ability to enter the vagina. The muscle contractions of orgasm propel the sperm forward and out of the penis. Since the sperm come out the same opening as urine, it's important to block off the bladder before the sperm get released. The erection helps block the urethra (the tube from the bladder). In other words, you can't "come" and "go" at the same time. This also adds to the pressure pushing the sperm out.

About two hundred million sperm are ejaculated into the vagina. And then the journey continues through the cervix, the uterus, and into the oviducts, where fertilization occurs. Fluid flowing from the uterus gives the sperm direction, and the sperm "swim upstream," monitoring the currents from of these fluids. In humans, there does not appear to be much "competition" among sperm, as they're usually all from the same person. In some species of mice and in many insects, where a female can have several reproductive partners within a few minutes, there is competition, and the sperm swim faster (Edwards et al. 2014).

But the female reproductive tract is not merely a racetrack for the sperm, and when the sperm get into the oviduct, something incredible happens. The oviduct cells extend membrane processes that wrap around the sperm and hold them there. Sperm don't race toward the egg through a passive tube. Rather, the oviduct cells hold the sperm tightly. That's because the sperm can't fertilize the egg yet! Their cell membranes aren't mature enough to fuse with the egg's cell membrane, and they can't sense the presence of the egg yet. In other words, when a sperm is ejaculated, it can't fertilize the egg. They need to be matured by the oviduct cells through a process called capacitation. Capacitation, the acquisition of ability to fertilize an egg, is accomplished by the cells of the oviduct holding the sperm and changing its cell membrane. Only when the sperm are released can they sense the egg and fuse with it (Cohen-Dayag et al. 1995).

Capacitation is a crucial step in fertilization, but is often left out of popular stories of sperm racing toward the oocyte. Most animals do not have this capacitation step. Frog and fish sperm, for instance, are capable of fertilizing the egg as soon as they are released. The recognition of a need for capacitation in human fertilization and the research into how it occurs were key events in the development of IVF procedures to combat human infertility: IVF includes a step in which the sperm are artificially capacitated (Chang 1951).

FERTILIZATION: THE SPERM AND EGG MEET EACH OTHER

Of the millions of sperm ejaculated, less than a dozen make it to the area around the egg. The area where the egg resides is a little warmer than the other parts of the oviduct, so the sperm follow the heat to where the egg is. Then, hormones (such as progesterone) from the cells surrounding the egg attract the sperm to it.

Those sperm that have been capacitated recognize these signals and lash their tails to travel toward the egg. They also open up the sac of digestive enzymes in the acrosome at the tip of the sperm. The combination of enzymes and rapid tail movement get the sperm through the weakly bound cells that surround the egg (the corona radiata). There, the sperm meets the proteins that encircle the egg cell. This protein coat is called the zona pellucida (which means "transparent belt" in Italian), and it has two main functions. First, the zona pellucida proteins serve to recognize the sperm. It's like a secret handshake, and it makes certain that the sperm and the egg are from the same species. The zona proteins bind the sperm to the egg and help guide the sperm toward the egg cell's membrane. Second, the zona pellucida prevents the egg (and the early embryo) from binding to the oviduct cells, as the egg must adhere only to the womb. If the embryo "hatches" from the zona too early, when it is still in the oviduct, the embryo will try to implant into the oviduct as if the oviduct were a uterus. But, unlike the uterus, the oviduct is not prepared to support a pregnancy. Thus, an ectopic (tubal) pregnancy can cause bleeding around the embryo, potentially leading to a hemorrhage severe enough to cause the death of the woman.

As we saw while explaining why is it that we seriously need to arm ourselves with solid defenses against dark arts, one of the most enduring clichés concerning fertilization is that of the sperm powerfully entering the egg while the egg just sits there begging to be entered and begin the process of conception. Again, a fertilization process of this sort just doesn't exist. Therefore, let's once more revisit the process, its false descriptions, and its much less violent reality. Once the sperm touches the egg, the sperm does not "drill," "bore," or "plunge" through the egg. Rather, the sperm head, containing the nucleus, "spoons" with the egg.

The curvature of the sperm head matches the curvature of the egg. So the sperm lies next to the long-sought-after egg. And then, their membranes fuse, making them one (Satouh et al. 2012) (figure 3.5). The sperm and egg melt together, and the sperm nucleus finds itself within the enormous cytoplasm of the egg.

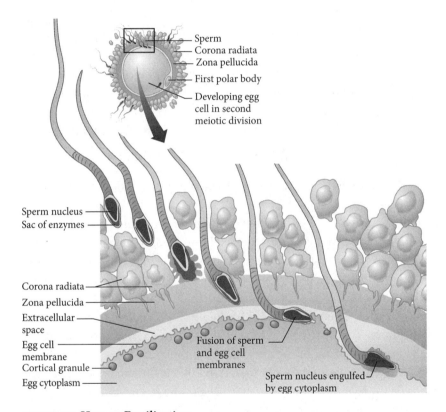

FIGURE 3.5 Human Fertilization.

When a human sperm is capacitated, its tail is given more energy to move, the enzyme sac breaks (allowing it to go through the cells surrounding the egg), and it can now adhere to the zona pellucida surrounding the egg. Once getting through the proteins of the zona pellucida, it turns so that its membrane is lined up with the membrane of the egg. The two membranes fuse, and the sperm cytoplasm is now contained within the egg cytoplasm. The nucleus, tail, and proteins of the sperm are now all inside the egg.

But once one sperm fuses with the egg, it is absolutely critical that no other sperm be able to do so. If two sperm were to enter the egg, the fertilized egg would have sixty-nine chromosomes instead of forty-six. It would also be told to divide into four cells rather than two (since each sperm brings in the proteins needed for cell division.) In such cases, different numbers and types of chromosomes are given to each cell, and the cells soon die. Therefore, it is critical that the egg have a way of making sure that one, and only one, sperm enters. And the egg has such a mechanism. Right next to the cell membrane of the egg are little packets of digestive proteins (similar to the single enzyme packet of the sperm), and one of the proteins in this packet digests the protein that binds the sperm to the zona pellucida. When the first sperm enters the egg, these packets all release their enzymes. When these enzymes alter the zona protein that binds sperm, further sperm cannot reach the egg. So, once a single sperm has entered into the egg, the egg gets rid of all other sperm.

Thus, a single sperm and a single egg unite. The entire sperm, tail and all, goes into the egg. The proteins that had once been inside the sperm are now inside the egg, and they activate certain proteins inside the egg (Gilbert and Barresi 2016; Ducibella and Fissore 2008). These newly activated proteins start the development of the embryo, one of the first actions of which is to get the egg to finish its meiotic divisions. One of the products of this division is a polar body (a small cell that will soon degenerate), whereas the other haploid nucleus remains inside the egg. The nucleus of the sperm and the nucleus of the egg, each containing half the normal number of chromosomes, then migrate toward each other to make the nucleus of the embryo. As we will see, the ability of the proteins inside the sperm to interact with the egg proteins to initiate development are the basis for intracytoplasmic sperm injection (ICSI)—the ability to initiate fertilization by injecting a sperm into an oocyte. Moreover, the chemical reactions by which the sperm activates the egg are almost identical to those used by the oviduct to activate the sperm. Reciprocity is the name of the game.

The sperm and egg are amazing cells. They take the father's genes and mix them with the mother's genes. Thus, from a genetic point of view, you are half your mother, half your father. And that means you are also getting a quarter of your genes from each grandparent. Indeed, the precursors of our sperm and eggs descend from cells that used to be in the gonads of reptiles, amphibians, and fish. Each of us is the lucky result of a lineage going back to the origins of life.

When the fertilized egg is formed, it is about the size of a sand grain. It is also referred to as a zygote or one-cell embryo. The cells of the oviduct then ripple to gently move the zygote toward the uterus. While traveling toward the uterus, the zygote divides several times, creating a ball of cells. At about the sixteen-cell stage, some cells are on the outside of this ball, and some cells are on the inside. This is a critical moment: The cells on the inside will become the embryo; these form the embryonic stem cells. The cells on the outside will become part of the placenta; these outer cells are critical for attaching to the uterus. They make adhesion proteins that bind like Velcro to similar adhesion proteins on the lining of the uterus (the endometrium).

The cells start producing a fluid that swells the embryo, making the outer cells touch the zona pellucida and push against it. At this stage, the embryo is called a blastocyst, which is a fluid-filled ring of cells, with one end containing the embryonic stem cells. And just as the embryo is about to enter the womb, the outer cells of the blastocyst make an enzyme that digests the zona pellucida. The zona had been critical in preventing the embryo from adhering to the oviduct. Now, however, the zona must be removed so that the embryo can attach to the uterus and receive the physical support, nutrition, and oxygen needed for development. When the pressure of the fluid pushes the blastocyst out of the zona, the embryo is said to have "hatched."

Right now, the important cells are the outer ones. The inner, embryonic stem cells, will form the body, but the task at hand is for the outer cells to bind to the uterus and stay inside the mother's body. The attachment of the blastocyst to the uterus is often called implantation, since the blastocyst will burrow into the uterus. Once the blastocyst has started burrowing into the uterus, the woman is considered to be pregnant. Yes, the medical definition of pregnancy is when the woman is carrying the embryo in her uterus, not when fertilization takes place. By implanting into the uterus, the *embryo* makes a woman pregnant. A man does not "impregnate" a woman. That's just another spell.

SPELLS AND MISCONCEPTIONS

By now we should have jettisoned a good number of spells about fertilization. The first sperm that reaches the egg doesn't necessarily fertilize it, since sperm that haven't undergone capacitation aren't capable of binding to the zona pellucida. Indeed, the female reproductive tract (consisting of the cervix, uterus, and oviducts) is not just a passive race course, and the egg isn't just a passive cell that sits there waiting for an active sperm. Rather, the female reproductive tract capacitates the sperm, and the egg attracts it. After the first sperm enters, the egg actively gets rid of other sperm. So both the sperm and the egg are active participants in fertilization. Moreover, fertilization isn't an aggressive drilling of the egg by the sperm. Rather, it's a fusion of the cell membranes, allowing the two cells to lose their individual identities and become one with each other.

However, as was mentioned in chapter 1, the media is still full of stories concerning the active sperm and the passive egg (BGSG et al. 1988; Martin 1991), and the story of sperm and egg often becomes one of human courtship or militaristic fantasy in which a sperm "captures" a fortified egg.

As we have shown, these stories have nothing to do with the cellular biology of fertilization. They are myths. They are spells cast on us to make us think of fertilization as some sort of competitive story in which you and I are the products of a successful warrior and his prize. Glorifying warfare and male competition, they leave out the facts that the oviduct activates the sperm, that the egg and sperm membranes fuse together, and that the egg is an active partner in fertilization.

It is important to base our policies on real science, not myths or wishful thinking. It is also important to know and understand the science, because what we think of nature determines who we think we are. Writes the religious philosopher A. J. Heschel (1965), "A theory about the stars never becomes a part of the being of the stars. A theory about man enters his consciousness, determines his self-understanding, and modifies his very existence. The image of man affects the nature of man." The view of fertilization as solely competitive and aggressive makes such behavior seem normal. What science finds, however, is that fertilization is to a large degree a remarkable cooperative effort among sperm, egg, man, and woman.

INFERTILITY

Now that we know some things about fertilization, we can look at what causes some couples to be infertile. Infertility is usually defined as the inability to become pregnant despite having frequent intercourse for over a year (CDC 2014). About 6.7 million American women aged 15 to 44 years have impaired fertility. Approximately 85 percent of young couples will conceive over the course of a year. Another 10 percent will usually conceive within two years. Fertility decreases with the age of the woman, such that it is 86 percent at ages 20 to 24 years and only 63 percent for women aged 30 to 34 years. In young couples, about one-third of the time, the cause of infertility resides in the man; about one-third of the time, the cause of infertility resides in the woman; and about one-third of the time, the cause of infertility is unknown (CDC 2014).

Male infertility can be due to low sperm production, failure to deliver the sperm properly, or the production of poor-quality sperm. Low sperm production can be caused by genetic mutations or by health conditions such as mumps, diabetes, or undescended testicles. Radiation and some toxic chemical exposures also cause low sperm production. Sperm-delivery problems can be caused by erectile dysfunction or by the physical blockage of sperm. Some genetic diseases (such as cystic fibrosis) can cause impairment of the sperm ducts, and in some instances, genetic mutations affect sperm motility or the ability of the sperm to fuse with the egg.

Female infertility can be caused by low egg production, failure in the delivery of the egg, and poor egg quality. In addition, blockage of the female reproductive tract, menstrual disorders, and even immune system abnormalities can cause female infertility. A failure of egg production can be caused by endocrine disorders such as polycystic ovary syndrome, thyroid dysfunction, endometriosis, or the overproduction of prolactin (the hormone that allows milk production). Excessive weight loss can also stop ovulation. Any disease that blocks the female reproductive tract (such as infection, inflammation of the oviduct, uterine fibroids, or benign tumors of the cervix) can prevent the sperm from getting further into the oviduct and thereby lead to infertility. In some cases, menopause occurs earlier than normal, stopping the maturation of oocytes. In addition, a woman's immune system can sometimes attack the sperm, preventing their arrival in the oviduct.

Environmental chemicals have also been linked to infertility. Smoking increases the risk of infertility in both men and women, and certain chemicals, such as bisphenol A (BPA), have been found to induce oocyte defects in rhesus monkeys (Hunt et al. 2012). Further, humans with high exposure to BPA have been found to have statistically fewer offspring than those with lower exposure (Caserta et al 2013; Lathi et al. 2014). Obesity lowers fertility in men and women, and sexually transmitted diseases, such as chlamydia, can damage the oviducts. Even mental stress can inhibit sperm production and ovulation.

But for about one in every three couples, the cause of infertility cannot be identified. We certainly do not know everything we need to know about infertility. Experimentation cannot be done on humans, and we often rely on medical problems to tell us what we need to learn. Despite our significant advances in knowledge, our understanding of fertilization and implantation is still at a very early stage.

4

FERTILITY RITES

———

Artificial Insemination and In Vitro Fertilization— Their Hopes and Their Fears

CLARA PINTO-CORREIA

Whenever God grants us a gift, he also always hands us a whip.

—Truman Capote, *Music for Chameleons*

Fertilization is what takes place after intercourse. A sperm meets the egg, the embryo develops, and nine months later, the cradle is filled with a baby. And they all live happily ever after.

However, for many couples, this fairy tale has never been the case. Infertility is not an uncommon affliction in humans. Although these values are never easy to settle, most recent studies agree that nearly 15 percent of couples worldwide cannot conceive a child (Volpe 1987). In 2010, 11 percent (7.4 million) of all American women aged fifteen to forty-four experienced "impaired fecundity" (yet another euphemism for infertility), and 6 percent of married American women were diagnosed as infertile (Roberts 2006; Wilson 2014). Worldwide, over 75 million people are reported to be "involuntarily childless" (Wilson 2014). The issue has lately gotten so quantified that the abundance of figures and estimates becomes almost numbing.

Medicine and agriculture have long been inventing technologies to alleviate infertility in humans and livestock. But technologies never exist in social vacuums, and if they cure old problems, they have the potential to create new ones. In a multinational, multi-religious, multicultural community, with fewer and fewer boundaries between peoples and places, who or what is to decide if a new technique is a gift of the heavens or a curse from hell? What started out in the West as a set of innocent techniques to ameliorate farmers' lives and improve livestock production became a complete set of human medical practices that, in a historical split-second, profoundly changed our concepts of life and of family. Were we ever ready for this?

To a large degree, our response has been framed according to our expectations and our anxieties. During the second half of the twentieth century we suddenly acquired great expectations that had never been a possibility before. DDT could exterminate thousands of pests, and its capacity to fight malaria and typhus saved thousands of lives during World War II. Likewise, the heroic era of medical microbiology—the identification of the bacterial agents of infectious diseases and the making of antibiotics and vaccines to eliminate them—gave us the idea of a future of boundless control over nature. The Western mind ripened to the notion that science would soon acquire unlimited powers over *all* our bodily afflictions. And this, as clearly expressed in the period's science fiction, would certainly be extended to one of the most intimate spheres of our lives—the making of babies and the invention of new ways to gain control over what used to be totally out of our reach. Historically and culturally, we can easily read the eager scientific response to the first breakthroughs in assisted reproductive technologies (ART) as the logical consequence of the growing expectation for more and more medical miracles that would set us free from the limitations and impositions of nature.

ARTIFICIAL INSEMINATION

For the first centuries of their use, however, and long before science fiction existed, ART enjoyed a noncontroversial existence. The first technique to enter the real-life scene, as was to be expected given that it is the simplest of them all, was artificial insemination (AI). It basically consists of recovering sperm from the male, and introducing it into the uterus or cervix of an ovulating female. Today, one can artificially concentrate sperm in very fine straws, freeze them at very low temperatures, and finally thaw them by the unit when needed. The key element of AI is so easy to carry out (collecting the semen from a male and inseminate an ovulating female) that its first scientifically detailed and reported success dates back from the eighteenth century. However, the idea of AI has a much longer and more interesting history.

The first AI attempt that we know of involves King Henry IV of Castile, who was in a second marriage owing to a fruitless quest for an heir with his first wife. Continuing to struggle with infertility, he finally decided it was appropriate to seek the help of the court's physician. Ingeniously, this fifteenth-century gentleman invented a syringe-like device with a reed made of gold and used it to collect a sample of Henry's royal semen and transfer it into the queen's vagina.[1] No questions asked, now—the procedure worked. Alas, the baby born from this artificial insemination was a girl, who, because of her gender, never got to reign after Henry's death, as often happened in these cases. Rather, she was substituted on the throne by the famous Isabella the Catholic (Munzer 1924), who married Ferdinand of Aragon, unified Spain, created the Inquisition, patronized Columbus, expelled the Jews and Muslims from Spain and Portugal, and thus completely changed the world.

Three centuries later, AI was performed with astonishing success in dogs by the Italian Catholic priest and great microscopist Lazzaro Spallanzani in 1782, leading to the birth of eight happy spaniel puppies (Pinto-Correia 1997). This success immediately encouraged farmers to start their own experiments with cattle, mainly with the aim of inseminating as many good cows as possible with concentrated semen of one single above-average male, thus considerably speeding up the rate of genetic improvement in animals destined for producing meat and milk.

Next in line would be fearless human beings. Some nineteenth-century American women were brave enough to allow their doctors to use all means possible to get them pregnant, regardless of the pain these doctors could cause them. During the 1870s, using a brand-new instrument of his own design, controversial surgeon J. Marion Sims, one of the heads of the Women's Hospital at Philadelphia, spent two years working with a half-dozen patients and their husbands, performing a total of fifty-five artificial inseminations—and had only one case of success, which ended in a miscarriage. Sims had failed to take the timing of ovulation into account (Marsh and Ronner 1996).

However, about a decade later, also in Philadelphia, physician William Pancoast was counseling a woman about her inability to conceive. He took the time to determine that the woman could be fertile but that her husband had a low number of sperm. It seemed that an earlier case of gonorrhea had left his semen void. Telling the woman that he needed to perform another examination of her body, Pancoast sedated the patient with chloroform and inseminated her with a rubber syringe, containing the sperm of a medical student whom he had deemed the most attractive. (Informed consent was not something that existed until after World War II). Nine months later, the woman gave birth to a healthy boy. The husband, but not the patient herself, was told the circumstances of the birth (Yuko 2016). The medical students were sworn to secrecy.[2]

While advances in humans seemed slow and secretive, results in animals couldn't have been more promising or public. In the last years of the nineteenth century, Cambridge-based reproductive biologist Walter Heape and Russian biologist Ilya Ivanovich Ivanov reported success in rabbits, dogs, poultry, and horses through the use of AI. These innovations inspired the organization of AI cooperative dairies in the United States and Europe. The artificial insemination of cattle was certainly an extremely smart move, since it is obviously much cheaper to keep hundreds of straws in liquid nitrogen, where they seem to this day to remain usable forever, than to keep stables and pastures full of bulls, rams, or stallions that require extreme care to be maintained and besides are ungrateful enough to eventually die. Also, this technique generally comes with optimal results. By the mid-1980s, rates of farm animals born through AI per Western country were steadily located around the 50 percent mark. More recently, more than 90 percent of dairy cows were reported to have been artificially inseminated in the Netherlands, Denmark, and the United Kingdom (Ombelet and Van Robays 2015).

The Fast-Spreading First Human Artificial Inseminations

There was no quantum leap from inseminating animals to inseminating people. Indeed, it was expected that what worked in domesticated species would soon work in humans. This view was so pervasive that it became a crucial plot device in D. H. Lawrence's *Lady Chatterley's Lover*. Such medical miracles were assumed; *Lady Chatterley's Lover* was published in 1928, before the first successful scientific reports of human AI were published—but it was not considered science fiction. Actually, the first public reports of successful human AI originated from Guttmacher (1943), Stoughton (1948), and Kohlberg (1953a, 1953b). By the mid-1980s, the total number of children known to have been born through AI was already amounting to over a quarter-million. A substantial number of those children had been brought to life through the services of human sperm banks, public and private alike, which blossomed from the early 1970s onward. In the United Kingdom, for instance, by 1980, around four thousand babies were being born yearly by these means (Pinto-Correia 1986). A totally new set of moral and social problems had just been created, but nobody was all that aware of them just yet.

Banking Sperm

Sperm banks posed several ethical questions as soon as they appeared, but their existence and practices were, and still are, largely unregulated. Instead of coming to terms with a simple core of basic universal guidelines, society has rather been increasingly left to face a dramatic kaleidoscope of values and interests. To start with, some of these sperm banks are public whereas others are private. The public sperm banks have had long waiting lists since they first began operating. Private sperm banks promptly started specializing in such perks as Nobel Prize winners' sperm or Scandinavian-only sperm, with prices elevated according to market demand. These dichotomies would be problematic enough themselves, even if nothing else had changed in the world of assisted reproduction since these early days.

The moment you introduce the concept of selling "better sperm" for a higher price, you're faced with an immediately ensuing controversy that might well be a recurrent theme for the remaining chapters: Are such dubious offers to be tolerated at all, since, according to their premises and their cost, they are, by definition, *biologically* separating rich from poor? At many campuses, sperm donors are recruited among intellectually, cosmetically, or athletically superior students, and several companies publish catalogues of "plus" donors of all sorts, telling women about the physical, mental, and psychological attributes of possible fathers for their children. This leads straight to the second obvious problem: all these promises of "plus" sperm are highly debatable at best. We all know that genes play their own games and that some may remain suppressed during a considerable number of reproduction cycles before reappearing in a subsequent generation.[3] Also, "plus" sperm is by no means all it takes to have a "plus" child. Dancer Isadora Duncan is said to have proposed to playwright Bernard Shaw that they have a child together so that it could have her radiant looks and his brilliant brains. Shaw reportedly refused this tempting offer, explaining that the child could end up having *his* looks and *her* brains. Moreover, as has been discussed earlier, what the fetus experiences inside the uterus and what the newborn experiences during the first years of life are critical in forming traits and behaviors, too. Besides, a trait or behavior that is "favorable" in one environment might not be so in another. Prospective clients have never been expected to be knowledgeable on the topic of genetics, and the owners of the sperm banks often believe that the public will accept that quality men give quality sperm to produce quality babies.

And how about the donors? Is it acceptable that they are paid for their "services"? Should anonymity be a mandatory requirement, for the sake of everyone's peace of mind? Or should the identity of the biological father be revealed upon the child's request at age 18—or even earlier, depending on what a given culture may decide is best for ensuring self-identity or family harmony? Moreover, how about the risks of inbreeding in cases of small communities with one very active donor? It has already happened. In two famous cases, physicians were found to have artificially inseminated women with their own sperm, including British physician Bertold Wiesner, who is said to have been the sperm donor for six hundred women (*New York Times* 1992; Smith 2012). Inbreeding can be dangerous, as members of the Habsburg royal family remind us (Pinto-Correia 2003). There's a reason churches once posted wedding banns and officiants asked if there was anyone who knew why a couple should not be wed.

Modern Eugenics

Since unregulated private sperm banks are still offering "premium choices" of all sorts, the situation is less and less reassuring as time passes. By now, the techniques that recognize small changes in DNA (technically "single-nucleotide polymorphisms," or SNPs, pronounced "snips") are getting progressively better and cheaper. As it happens, modern SNP research, as abundantly financed as it has been so far, stands a worrying chance of being used in order to offer the richest clients of the most sophisticated sperm banks the "scientific" promise of intelligence, athletic prowess, or physical beauty for their children. The selective capacity of SNPs lies in the fact that they can act as molecular markers to identify the presence of a great number of genetic features in a single sperm sample. As happened with a great number of assisted reproductive technologies, SNPs were originally a great medical idea, initially studied and developed to identify disorders such as hemophilia or cystic fibrosis. Now, it's claimed that SNPs can recognize the genetic substrates for desired human traits. There is, for example, a SNP that can recognize the gene responsible for blond hair. There is a SNP that is known to give people large muscles. Are we going to give the wealthy the ability to order the traits of their offspring? Are techniques such as clustered regularly interspaced short palindromic repeats (CRISPR; discussed in chapter 7) going to be used to mutate genes to produce whatever combinations people can pay for? Are there going to be fads in baby characteristics just as there are in baby names?

The term *eugenics* is derived from the Greek for "well-born," and the concept has been discussed throughout history, from the days of Sparta and Plato's *Republic*, in the medical projects of French Illuminists, and even in the dreams of the French Revolution: the goal is to breed better-built, healthier people with each other so that the human race becomes more perfect with each successive generation. The concept finally got its scientific name and its complete program in the writings of Francis Galton, the childless half-cousin of Charles Darwin, who felt that the breeding techniques used in farming could be applied to humans. According to Galton, the best of humanity should be encouraged to reproduce, whereas the weak (in either mind or body) should be placed into the secular equivalent of monasteries and convents, where they would receive care but not "burden" the next generation with their descendants.

These ideas were fervently endorsed in the United States by Charles Davenport and Harry Laughlin, the founding fathers of a eugenics movement that spawned an amazing number of societies, associations, science institutes—and even landmark traditions, such as the annual contests "America's Best Baby" and "America's Fitter Family." Following Davenport's concept that eugenics was meant to "improve the natural, physical, mental, and temperamental qualities of the human family," the movement managed to sterilize an amazing number of people against their will, centering its efforts particularly on "undesirable traits," such as pauperism, mental disability, dwarfism, a number of transmittable and venereal diseases, promiscuity, and criminality. In an attempt to privilege Northern European genes in the United States, Congress passed the Immigration Restriction Act of 1924, the first major legislation proposed to restrict the gene pool of the country to that of Northern Europeans (Ludmerer 1972; Kevles D. J. 1998; Carlson 2001).

Hitler and his inner circle made it very clear that they had learned a lot from the strategies of the American eugenicists. Several prominent Americans in the eugenics movement boasted likewise. During the Nuremberg trials, some German eugenicists protested that they had done nothing different than follow the agenda proposed by the Americans (Kühl 1994). We know the rest of the story. And yet, here we are today, willing to allow people to buy themselves babies with the "best" traits. We act as if we really know what the best traits are and how to dependably select them. In Germany, eugenics was enforced by the government. In the United States, however, it was often said that eugenics would come about not by coercion, but by economics. And the first person to say that was likely the founding father of the Soviet Union Leon Trotsky in 1935.

Still, there are many reasons why these eugenic promises are not going to be fulfilled. They will be the cause of even more frustration for people aiming for a "perfect" baby. First of all, meiosis is the great trickster of evolution. If a person is an athlete, this athleticism may be the result of a combination of traits, including bone mass, muscle mass, tendon arrangement, capillary development, and the amount of red blood cells made per minute (to give oxygen to those muscles). A person whose genes give him these traits might have the potential for great athletic performance. But meiosis is going to jumble those genes up. The new baby is going to be a mix of some of the Father's genes and some of the Mother's genes, and the combination that made the Father so amazing might never be seen again. Only half of the new baby's genes will come from the "plus" Father, and it is extremely unlikely that all the genes that made this Father a "plus" will be in a single sperm. Indeed, as we will see, this recombination is the rationale for cloning mammals. The transgenic sheep with a great amount of human protein in their milk would die, and one could not be certain that the combination of genes that made those sheep so spectacular would ever be seen again. So the biologists tried to clone them.

The second limitation faced by eugenics is that a gene doesn't function alone, but rather within an ecosystem of interactions with the products of other genes. There are genes that produce a perfectly normal-looking face when mixed with certain combinations of other genes, but which lead to malformed faces when mixed with different ones. A normal-looking sperm donor may harbor mutant genes that will become expressed when combined with those of the egg his sperm happens to fertilize. The fact that the same gene can give rise to different appearances in different situations is called "phenotypic heterogeneity." In mice, the gene that leads to the formation of testes sometimes doesn't work in eggs having a different genetic background. Even in the age of SNPs, life still has plenty of uncertainty to throw our way (Gilbert 2002; Gilbert and Epel 2015).

And third, there is no gene for intelligence or athleticism or leadership or musical talent. There are sometimes some genes that are associated with such talents and may even contribute to them. There are certainly genes that can give one more neuronal plasticity, and people with these genes might learn faster. And there are some genes that are known to increase muscle mass, rendering a person more prone to be athletic. In 2014, National Public Radio ran a fascinating piece on "genius sperm," in which a man conceived by such sperm finds his father, only to be very disappointed. Fortunately, the sperm donor gives his biological son some good advice: He can go his own way, make his own choices, and not be like either his biological or social father (Washington 2014).

Most likely, we should be thankful for these shortcomings. But many scientists are fearing that, as our ability to identify genes becomes progressively better, sooner or later the day will come when we can accurately predict the outcome of combining a certain sperm with a certain egg. If we should get there, will we be able to make, if not "designer," at least "better" babies? Once more, will the wealthy be able to afford sophisticated techniques to ensure their babies are given the "best genes," while the less affluent will still have to rely on old-fashioned chance? Or will we go as far as truly having the by now infamous "designer babies," whose parents choose the particular characteristics they want their infants to have? This is not idle chat. Scientists as powerful as James Watson (2016), one of the discoverers of the DNA double-helix and a major proponent of germline modification, has said that "eugenics is sort of self-correcting your evolution. . . . I think it is irresponsible not to try and direct the evolution to produce a human being who would be an asset to the world as well as to himself." These issues have never really been debated on a vast, organized scale. They deserve international and organized attention now, rather than after "designer babies" are announced in our newspapers.

Good Intentions and Complex Outcomes

Obviously, however, AI is a godsend in several circumstances. This technique can serve a huge spectrum of sound medical purposes. It can save the sperm from a man about to undergo a vasectomy or some other treatment (such as those for cancer) that might make him infertile. It can allow the use of concentrated sperm if a man's sperm count is low or the use of sperm from men with impotency or premature ejaculation. Or it can even be used to inseminate a woman with the semen of an anonymous donor in cases of total male infertility springing from a host of medical reasons. Also, it easily bestows motherhood upon women who want to live on their own, totally dispensing with male partners.

Still, good intentions soon lead to complex moral headaches. One of the first examples hit Europe as early as 1974, when a young man was killed in a car accident. Soon afterward, his French widow demanded to be artificially inseminated with her deceased husband's frozen sperm (Pinto-Correia 1986). The public perceived this as a morbid request, but should this childless young woman have been respected in her medically feasible wish? Should a dead individual beget a living person? What will we tell the child? There was an immediately ensuing commotion, with equal amounts serious debate and morbid sensationalism—and finally, against most expectations, the court ruled in the woman's favor. However, all the ensuing AI attempts failed. The wife did not have *partum post mortem*. The questions are still out there to be settled. So much so that forty years have passed, and the area still remains troubled by many unexpected problems.

Many decades later, unresolved social and moral issues springing simply from the existence of sperm banks and AI can still split opinions and fire up arguments. In the absence of communal discussion and guidelines, many such cases end up in court. One recent court decision dealt with the issue of whether the lesbian partner of a woman impregnated through AI would be considered a "parent" if the couple were to separate. Another court decision dealt with a case in which the infertility clinic clerk goofed, giving a couple who desired a white baby the semen of a black man (Nelson 2014). The white couple sued the clinic for wrongful birth, and lost. They are now suing the clinic for incompetence, because they have had to move in order to give their baby the upbringing they feel she needs. A healthy birth is obviously no longer the inscrutable miracle it used to be. Our conceptions of conceptions are changing.

IN VITRO FERTILIZATION

Artificial insemination was developed to serve us, but it has also haunted us with its unpredictable social impacts after it began to be used (Ombelet and Van Robays 2015). Likewise, when scientists started focusing their research efforts on the promise of in vitro fertilization (IVF), the outcomes of the technique could not be more enticing for the future of human reproduction. Although infertility can be caused for many reasons (as mentioned in chapter 3 and the appendix), one of the most common is the blockage of one or both of the female oviducts, caused by illnesses such as pelvic infection, endometriosis, tumor growth, and so forth. Malfunctioning tubes account for at least 20 percent of the cases of infertility (Volpe 1987). Using IVF, these conditions become suddenly easy to manage: the wife's mature eggs can be retrieved from an ovary, fertilized in vitro ("in glass") with the husband's sperm, and the resulting embryos, if showing good quality, can be transferred directly to the uterus. Among other problems, IVF can also help those women who have failed to conceive because the chemical fluids in the cervical canal had deleterious effects on the sperm, or those couples for whom the problem was a low sperm count. A good number of medical teams were seriously on this trail right after AI became possible.

And on July 25, 1978, when the renowned British surgical team of Robert Edwards and Patrick Steptoe announced the birth of Louise Joy Brown to the world . . . well, the world took a step back and gasped. A five-pound, twelve-ounce "test-tube baby" had been born. The then thirty-year-old Lesley Brown and her husband John had tried in vain to have a child for nine years, since both of Lesley's tubes were blocked (Volpe 1987). Now science had just gotten the upper hand over nature. But was this received as really good news?

The Public Fear of Test-Tube Babies

At least to a certain extent, we can argue that the initial media confusion and hostility over Louise Brown's birth sprang from the fact that Edwards preferred to immediately reveal his feat to the media rather than first submitting it to a scientific journal. Whatever reasons moved him, the situation was pretty much a precursor of the diplomatic disaster of Dolly the sheep. The news of the first "test-tube baby" was just dumped onto journalists and audiences who obviously had not received enough information about fertility or reproduction to understand what IVF was all about. And, thus, the media circus was relentless. It seems the only preparation had been dystopian science fiction books such as Aldous Huxley's *Brave New World*. Babies created outside their mothers' bodies for total control of their developing features according to the wishes of the ruling class—and there you have it.

This tension clearly showed right away in clinics and families alike. As described by sociologist Karen Throsby (2004, 4),

> Pioneers of IVF were so afraid of the effects of public perceptions of the procedure if the first baby was "abnormal" that the couples undergoing the first experimental cycles had to agree on an abortion if the developing fetus was discovered to be malformed. As an example of the continuing suspicion that IVF children will be somehow marked by their unconventional beginnings . . . [a child's mother] discovered to her horror that behind her back her son was referred to as Damien (the Antichrist in

the *Omen* series of horror films) among certain members of her family because he was conceived through IVF.

Philosophers and theologians obviously also had a say in these matters, as happened when Paul Ramsey of Princeton University, a conservative Protestant theologian, contended that IVF was an immoral form of pro-creation, since it should not be a goal of medicine to enable women to become pregnant artificially or to interfere with natural fecundity (Volpe 1987). Various religious groups chimed in, saying whether or not repro-duction had to be "natural" and whether IVF reflected lack of love.

The Public Dream of Miracle Babies

However, empty cradles being the powerful social force that they certainly are, reactions to IVF were quite likely one of the most vivid double stan-dards of the twentieth century. While the press ran scary headlines, and absolutely unprepared commentators foresaw a world disaster, couples with infertility problems were already lining up at the door of all the clinics offer-ing IVF treatments. Literally. At that time, even when they were in good hands, these people were plainly expecting miracles. As Christo Zouves and Julie Sullivan (1999, 19) state, "In 1985, success rates for IVF patients were about 10 percent. Ninety percent of the women completed a cycle without delivering a baby. Even when all the variables seemed right, more often than not the procedure failed. When it worked, we thanked God."

Trial and error have consistently played a major part in the develop-ment and perfection of reproductive techniques, from culture media to surgical hardware. This is how scientific research often works, but gener-ally it does so with mice or guinea pigs. But this time, the guinea pigs have systematically been human patients and their much-desired offspring. Does this scare them away? No, and it never did. From the very begin-ning, couples entering the universe of IVF cycles have shown an uncanny stubbornness when it comes to never leaving the game until something works, no matter the price, no matter the odds, no matter the nausea, no matter the risks of the procedures needed to attain their own "biological legacy." This is one of the chief reasons the industry of making babies has raised so many difficult questions over the decades.

Superovulation and Embryo Freezing

As is easy to imagine, right after 1978, private infertility clinics intent on becoming the best in their field—and being paid accordingly—sprang up just about everywhere. Originally, one fertilized egg would develop into a four-to-eight-cell embryo and be transferred into a woman's uterus. However, as in nature, the odds of the fertilization succeeding and the embryo nesting in the right place in the uterus were initially quite small. Therefore, to increase the chances of success, clinics started using hormones to cause the ovaries of women undergoing IVF to release many eggs at the same time (instead of the usual one egg), so that several eggs could be fertilized simultaneously and implanted during that cycle. This "superovulation" substantially raised the odds that a pregnancy would occur. Again, this was only one baby step to take from where the first techniques had left off, and IVF clinics worldwide were quick to use it.

Superovulation, pioneered by Drs. Howard and Georgeanna Jones at the Eastern Virginia Medical School, allowed couples to have children who would never had been able to bear them before. But this blessing brought with it unforeseen problems. The hormones used to stimulate the ovaries (follicle-stimulating hormone and human chorionic gonadotropin, which has an action similar to luteinizing hormone, as described in chapter 3) can upset the body and produce ovarian hyperstimulation syndrome. As the ovaries rapidly enlarge, a woman can develop a sickness that includes nausea, vomiting, and stomach cramps. That's the mild form. The severe form causes persistent nausea and vomiting, dizziness, severe abdominal pain, rapid weight gain, and respiratory distress that can require hospitalization. Fortunately, most cases are mild. Still, about one out of every twenty women treated with these hormones develops mild to severe ovarian hyperstimulation syndrome (Mayo Clinic 2014).

But hyperstimulation was just the first step of getting many potential embryos. Another step was to actually transfer into the uterus many of the embryos that did develop in the petri dish. By the early 1990s, physicians had enough embryos at their disposal to allow for the transfer of four, five, or even six embryos to a woman's womb. Some patients still had no luck.

But others started bearing twins, triplets, then even higher multiples as hormone treatments progressed.

This could be very risky because human pregnancies are optimized for a single birth at a time. Even twin pregnancies have much greater medical risk for both the mother and the babies. The situation gets even riskier with triplets, and much worse as people attempt to carry greater numbers of babies at one time. As will be discussed in later chapters, twins and triplets are often born too early and experience complications associated with premature deliveries. Such complications, in which the infants need medical care during the first days of their lives, occur in about half of all twin births and in nearly all triplet deliveries. Cerebral palsy, which may be a complication of prematurity and low birth weight, is also much more common in twins and triplets than in singleton births. About 11 percent of children with cerebral palsy are twins (whereas twins account for less than 2 percent of the population) (ACPRG 2013).

Another unanticipated problem arising from superovulation has to do with the rapidly adopted plan to freeze those embryos that looked viable but were not needed for the first attempt. The source of inspiration was, obviously, the frozen sperm used for AI and its impressive resilience. Attempts to freeze four-to-eight-cell embryos began in the early 1980s and met with truly encouraging success: They could be held perfectly in liquid nitrogen for long periods of time, and, upon thawing, they recovered all their functions and were perfectly capable of— with luck, right in the first cycle—nesting in the mother's womb and initiating a pregnancy. After a baby was born in Melbourne, Australia, from an embryo previously frozen in the mid-1980s, the success was confirmed in cities throughout the world. Moreover, these embryos even proved able to be frozen again if not used in the second cycle and could remain frozen until needed. This advance would allow IVF clinics with nitrogen coolers to save all the supernumerary embryos from each woman seeking their services and then use them in later cycles if the first didn't work, without having to put the patient through another round of harsh hormonal treatments to trigger superovulation. It was a fine idea.

With strings attached, like all others.

Controversies with supernumerary embryos arose immediately after the technique appeared. The story that follows was only the first one to rock the world, questioning things that no one had thought would ever have to be questioned. This happened simply because of an IVF procedure and its resulting frozen embryos—and it shook everything from legal matters to embryonic rights.

EXTREME EXAMPLES

Frozen Assets

Once upon a time in 1982, Elsa and Mario Rios were a millionaire Los Angeles–residing childless Chilean couple, undergoing infertility treatment at the Queen Victoria Medical Center in Melbourne, Australia. At their first round of IVF, they obtained three healthy-looking embryos, one for immediate transfer and two for freezing just in case. After the first failed attempt, the couple were advised to take time off and relax since their state of anxiety was not ideal for the procedure. They went on a vacation, but both died in a plane crash—without ever having made a will. So for the first time in history, there were two frozen "orphaned" embryos. With a lot of money at stake to be inherited, a long battle followed. What should be the human status of these two four-cell entities? Who's to inherit the fortune? Which country's laws are to be applied in this case: Chile's or Australia's—or California's, for that matter, since that's where the Rioses were residents? Two years of social and legal nightmares unfolded, and the final court decision taken in 1984 to simply offer the embryos for "adoption" to interested childless couples still has many critics. Relatives of the deceased couple struggled endlessly for their right to access the fortune, also frozen, in a Chilean bank (Pinto-Correia 1986; Volpe 1987, 60).

Legal Perplexities

Decades passed, and court cases piled up. Hundreds, if not thousands, of embryo lawsuits are currently being fought, including the much-commented-upon 2014 case of the man who wanted to start a family with the embryos he created with his previous fiancée, actress Sofia Vergara.

In cases like this, there are still no real rules. Legal cases have been brought by women and men alike, with mixed results depending on the state and the judge. Frozen embryos are not even all equal before the law, although their entrance into the legal system dates from 1982. Legislators and courts are as perplexed as everybody else before matters of this sort. Collectively or personally, we often just don't know what to do.

Too Many Embryos

Frozen embryos, however, have piled up even more than embryo lawsuit court cases. By 1991, tentative counts concluded that there were already *millions* of unclaimed frozen embryos stuck in a liquid nitrogen limbo worldwide. As often happens in these cases, the information presented in the media was dangerously misleading. Since most countries never came to a decision, some nations eventually followed England's lead and decided just to dispose of the now infamous supernumerary embryos altogether.

But the issue is obviously far from settled. In 1996, Pope John Paul II muddied the waters even further by claiming that scientists could be creating frozen human embryos on purpose so that they could later do whatever research they wanted with them.[4] This is the perfect example of how even a person with enormous responsibilities and numerous medical advisers can be grossly misled on very serious matters and how easy it is to send a chill through large communities of believers—including those who were directing IVF clinics then. Eventually, a good number of physicians decided to dispense with freezing embryos altogether. As obvious, this causes its own social problems: women needing more than one cycle would have to pay the financial and personal price of continual superovulation treatments.

Universal Issues

There is something to be said for the fact that IVF treatments have now expanded to places in the world where infection is so prevalent that a person can die from a simple surgical intervention for appendicitis. These often happen to be places where any husband can easily dispense with his wife if she proves unable to give him children within a short number of years.

An excellent series of anthropological studies carried out in Alexandria, Egypt, documents in detail what such women put themselves through trying their luck at private infertility clinics with dubious credentials, often combining these treatments with a number of not-so-harmless folk medicines (Inhorn 1994b, 1995). With the increasing publicity suggesting that medicine can now solve infertility problems, these women spend their money and ruin their health in a desperate quest to satisfy their family's demands, whereas the possibility of paternal infertility is hardly ever raised (Inhorn 2003). In a Muslim country ruled by a book stating "wealth and offspring are the adornment of earthly life," (Qu'ran 18:46), the thousands of women who are poor and infertile and thus cannot achieve the two main divine earthly blessings fail their husbands, their families, their relatives, their communities, and even their faith. They stand against theories of male procreation that predated Islam by several millennia and go all the way back at least to the Pharaonic tradition. Moreover, they lack the traditional source of power that women with children have, and they insult the frustrated fathers to whom the children are assumed to owe their lives (Inhorn 2003, 22). In situations like these, the fact that the techniques of ART are known (but inaccessible) is arguably more of a curse than a benefit in the complex workings of family life.

On the same note, we find an even worse situation in what is now casually called "the infertility belt": the vast areas of sub-Saharan Africa where the extreme harshness of life seems to render more and more women infertile at increasingly early ages. According to the few studies carried out in these regions, women are willing to walk for miles and sleep by themselves in the bush if need be to get to the next little town where it is rumored that an infertility clinic is in business—exhausted, hungry, dehydrated: never a good way to start a cycle. For most, these treatments fail and leave serious scars. But the demand is always high, because husbands can always get themselves another wife if they don't get all the children they want, even if the infertility problem is the husband's (Boerma and Mgalla 2001).

Accepting the Unacceptable

It can be moving to encounter a modern Muslim woman, assuming that had she only married a Westerner, the couple's families would never put them through so much grief (Inhorn 2015, 6). Western women know all too well that this is not the case, and they generally also know that the West has just about zero tolerance for scientific frustration. In 1996, Paulette Bates Alden told her own infertility story in a book called *Crossing the Moon*. At some point, after the description of several failures, we meet a doctor who seems to be a textbook illustration of how clueless biomedicine seems to find itself in the face of the mysteries of reproduction (Alden 1996). Alden had failed yet another attempt at getting pregnant through IVF and was going through yet another cycle. A few days after the embryo transfer, she soils her underwear and starts cramping—sure signs that the embryos did not implant. When she asks her doctor what to do, he momentarily loses it: "I don't know, go home, lay down, light candles, pray, surround yourself with fertility totems—for Christ's sake, just do what women have been doing for thousands of years!" Is he callous? Does he care too much and feel as frustrated as his patient? Perhaps it's that we face more unknowns than we probably would like to admit when it comes to our own or somebody else's desperate quest to elicit biological children from a body that was not physiologically meant to bear them.

Thus, it might make sense to hear some real words of wisdom. In 1958, Protestant theologian Paul Tillich (1952, 127) stated that "grace is accepting the unacceptable." If, after this long road of biomedical miracles, we become unable to accept infertility when our bodies cannot respond to all the smart tools we have invented to wake them up, this awesome new medicine has driven us backward, not forward. It wouldn't hurt us any to consider this possibility, too.

III

THE MOTHER AND HER FETUS

I n this section, we discuss the two main entities of pregnancy: the mother and the fetus. Chapter 5 looks at the beginnings of embryonic development and the function of embryonic stem cells. The development of the human embryo strikes both scientists and laypeople alike with awe and mystery. As with fertilization, organ formation involves interactions among cells. However, there is no consensus among scientists as to when the fetus becomes a "person." This chapter will discuss the various stages of embryonic development and why scientists reason that different stages might be the basis for "personhood."

Chapter 6 concerns how assisted reproductive technologies have dramatically altered our definitions of motherhood. We have invented new types of mothering and are just beginning to feel their effects on society. Surrogacy has a long history. Though it was often linked to prostitution, its practitioners are themselves often desperate to give others "the gift of life." Similarly, egg donation and postmenstrual pregnancies enable women to become biological mothers when such options had been unavailable until a few years ago. The new technology of egg freezing promises a life of freedom to young single women. These technologies are popularized and romanticized, but they are not without significant perils.

5

NORMAL DEVELOPMENT AND THE BEGINNING OF HUMAN LIFE

Why Scientists Are Being Asked Theological Questions
and Why Theologians Are Being Asked Scientific Questions

SCOTT GILBERT

The history of a man for the nine months preceding his birth would probably be more interesting, and contain events of far greater moment, than the three-score and ten years that follow it.

—Samuel Taylor Coleridge, "Notes on Sir Thomas Brown's *Religio Medici*"

In chapter 3, we saw how fertilization creates the beginnings of a new body. The zygote, the one-cell embryo, is barely visible to the naked eye, but it somehow generates a body five to seven feet tall, with a heart on the left side, a mouth and an anus at their appropriate places, and two (and only two) eyes that are facing forward and are always in the head. The knee is a marvel of muscles, tendons, ligaments, and lubricant-forming tissue, each of which grows at precisely the correct place and connects at the appropriate locations. It is truly an awe-inspiring process, and developmental biologists, those people privileged to study embryos, are people who are constantly being amazed.[1]

How does the zygote do it? First, this one-cell embryo has to undergo *growth*, in which the cells multiply to form the millions of new cells of the early embryo. This growth must be so well controlled that both sides of the face fit together, and both feet are the same size. Moreover, small details in cell division make us look more like our parents than our friends. Usually.

Then, these cells have to undergo *differentiation*, in which some cells will become blood cells, some gut cells, and some nerve cells, bone cells, and so forth. And these early embryonic cells have to undergo *morphogenesis*, the formation of ordered tissues and organs, wherein multiple different nerve cells become the brain, and the gut becomes divided into the esophagus, stomach, intestines, pancreas, and liver (see figure 1.1). The bones of our pelvis have to take a different shape than the bones of our skull. Growth, differentiation, and morphogenesis are the basic tasks of the zygote. These are the tasks of embryo formation: embryogenesis.

EMBRYONIC CELL CLEAVAGE

The first stage of embryogenesis is **cleavage**. Immediately after fertilization, the cells divide once every twelve to eighteen hours. Moreover, the chromosomes in those cells make more copies of themselves such that at every cell division, each cell gets the same genes. One of the most scary things about mammalian cleavage is that it must be synchronized with the migration of the embryo into the uterus. We mammals are unusual animals, since both fertilization and embryogenesis occur inside the mother (see figure 3.2).

As the embryo is being swept down to the uterus by a slow current of fluid, the cells divide, and the first differentiation event takes place. Shortly after the formation of the eight-cell embryo (that is, after the third cell division, around four days after fertilization), these loose cells suddenly huddle together, forming a compact ball of cells. This compact ball develops into the blastocyst, a fluid-filled ball, in which a small group of *internal* cells are surrounded by a larger group of *external* cells (see figure 1.1). Most of the descendants of the external cells will *not* produce any of the baby's structures. Rather, they will form part of the **placenta**. The first function of the placenta is to stick to the uterus so that the embryo will not fall out of the mother (Fleming 1987). This is why most of the cells of the embryo are placental at this stage. The important thing is to adhere to the uterus! Once combined with the mother's tissues, the placenta provides the blood that supplies oxygen and nutrients to the fetus. It also produces the hormones that cause the uterus to remain soft and retain the fetus as it grows. In addition, the placenta produces chemicals that block the mother's immune system so that the mother will not reject the embryo.

The embryo itself is derived from the descendants of the *inner* cells of the sixteen-cell ball. These cells generate the **embryonic stem cells** of the **inner cell mass**. These cells give rise to the entire embryo and its associated yolk sac, allantois (waste sac), and amnion (water sac). All our hundreds of cell types, all our organs, and even our sperm and egg cells, come from the embryonic stem cells that comprise the inner cell mass (Tarkowski et al. 2010; Evans and Kaufman 1981).

TWINS

Cleavage is the time when identical twins can form. Indeed, before the separation of the inner cell mass from the outer cells, each cell has the potential to form an entire embryo. If separated from the rest of the cells, they could do just that. Thus, these early embryonic cells are said to be **totipotent** (from the Latin for "capable of becoming anything"). Later, the outer cells can form only the placenta. However, each of the cells of the inner cell mass is not yet determined to become any specific type of cell, and which cell type each will become depends largely upon interactions with other cells. (The cells of the inner cell mass are said to be **pluripotent**—capable of forming many things.) In fact, if inner cell mass cells are separated, they can form complete twins.

This ability of early embryonic cells to remain totipotent was discovered by developmental biologists in the nineteenth century who were thoroughly amazed by their discoveries. Hans Driesch, for instance, separated the first four cells of a sea urchin embryo and found that each could form an entire embryo. We see such amazing regulation in human identical twins.

Human twins are classified into two major groups (figure 5.1): *monozygotic* (from the Greek for "one egg"), or "identical," twins and *dizygotic* ("two egg"), or "fraternal," twins. Fraternal twins are the result of two separate fertilization events and have separate and distinct genotypes.

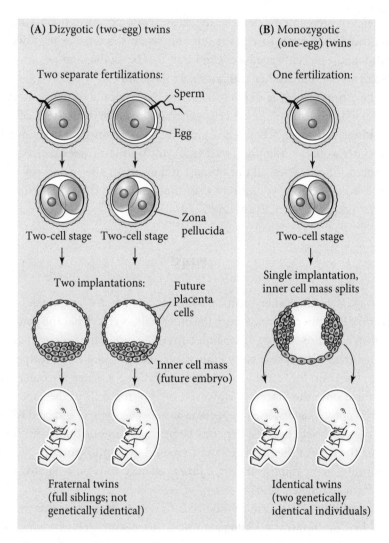

FIGURE 5.1 Twinning in Humans.

(*Left*) Dizygotic (two-egg) twins are formed by two separate fertilization events. Each embryo implants separately into the uterus, and the resulting twins are no more closely related genetically than any two full siblings. (*Right*) Monozygotic (one-egg) twins are formed from a single fertilization event. Sometime before day 14, the embryo splits in such a way that the cells of the inner cell mass are separated into two groups. Each inner cell mass group forms a complete fetus, resulting in two individuals with identical genomes.

(Thus, one twin could be male and the other female, or both twins could be the same sex). Identical twins are formed from a single embryo whose cells somehow dissociate from one another and thus share a common genotype. They are always the same sex. Identical twins may be produced by the separation of early embryonic cells, or even by the separation of the inner cell mass into two smaller clusters within the same blastocyst.

Identical twins occur in roughly 0.25 percent (i.e., one in four hundred) of human births. About 33 percent of identical twins have two complete and separate placentas, indicating that the split occurred before the separation of the outer cells from the inner cell mass at day five. The remaining two-thirds of identical twins share a common placenta, suggesting that the split occurred within the inner cell mass after the placenta formed. Multiple births are definitely more dangerous to maintain than singleton births, and the prevalence of congenital disease in twins and triplets is much greater than in the general population. As mentioned before, about 50 percent of twins are born prematurely (before 35 weeks) and at low birthweight, and this can lead to severe health problems, which can also translate into major financial difficulties for those who have to pay for medical treatments out of pocket. A recent American study (Lemos et al. 2013) found that, while a singleton birth costs about $21,500, the delivery of twins costs about $105,000, and the delivery of triplets around $400,000. In vitro fertilization births still account for a large percentage of multiple births in developed countries such as the United States, and the tendency for multiple births is rising, not shrinking.

HATCHING AND IMPLANTATION

When the embryo does reach the uterus, some five to six days after fertilization, it releases an enzyme that digests a hole in the zona pellucida, out of which the embryo hatches. Once out, the embryo interacts with the uterus, instructing it to prepare a dock for its attachment. The uterine cells respond by sending chemical signals to the embryo, instructing it to make the adhesion proteins that will bind to these docks. The embryo then makes direct contact with the **endometrium**, the inner surface of the uterus. The *endometrial cells* lining the inside of the uterus "catch" the embryo on a protein-containing "mat" that the endometrial cells secrete. This mat contains a sticky concoction of proteins that bind specifically to other proteins present on the embryo's outer cells, thus anchoring the embryo to the uterus (Wang and Dey 2006; Fritz et al. 2014). Once this "anchor" is in place, the outer cells secrete another set of enzymes that digest the endometrial protein mat, enabling the embryo to bury itself within the uterus. This process is called **implantation**, and it is the beginning of **pregnancy**.

At this point, a complicated dialogue begins between the outer, placenta-forming, cells of the embryo and the endometrial cells of the uterus. As mentioned in chapter 3, the placenta-forming cells "invade" the uterine tissue, secreting a hormone called **human chorionic gonadotropin** (the hormone measured in pregnancy tests). The human chorionic gonadotropin then instructs the ovaries to make another hormone, **progesterone**. Progesterone allows the uterus to remain soft and pliable, so that the embryo can grow, and it also prevents the uterine muscles from contracting, thereby preventing menstruation (which would eliminate the embryo). Finally, progesterone allows the blood vessels from the uterus to surround the embryo, and allows the uterine endometrium to expand to produce the decidua, the mother's portion of the placenta. Thus, the placenta is quite remarkable—a single, multifunctional, organ formed from two different organisms: the embryo and the mother.

Progesterone is obviously a very important chemical in maintaining pregnancy. In fact, chemically blocking progesterone's function in the uterus prohibits the embryo from implanting and thereby prevents pregnancy. This blockade is the method of action by which the drug **mifepristone**—sometimes called **RU486**—produces an early abortion (Chabbert-Buffet et al. 2005). It should be noted that this drug works very differently than (an in an opposite way from) the "morning-after pill" (discussed in chapter 3). The morning-after pill contains high levels of a synthetic progesterone that mimics pregnancy and blocks ovulation.

GASTRULATION

Now that the embryo is embedded inside the uterus, the inner cell mass can begin its own development. It separates into those cells that are going to form the amnion (water sac) and yolk sac and those cells that will form the body of the embryo. The cells that will form the body begin to become different from one another in a series of movements called **gastrulation**. Gastrulation (which originally meant "belly formation") begins about fourteen days after fertilization (right around the time of a woman's first missed period). It is during gastrulation that the embryonic cells lose their ability to be pluripotent. They can no longer regulate; that is, they cannot regenerate missing parts if some region of the embryo is removed. Thus, at gastrulation, the cells are given their basic instructions as to what they are to become. This also means that, at this point, twins can no longer form, and the embryo becomes committed to forming a single organism. This point is sometimes called "individuation."

Embryologist Lewis Wolpert (1983) famously (at least among embryologists) remarked, "It is not birth, marriage, or death, but gastrulation, which is truly the most important time in your life." This is because gastrulation is the time when the fates of the embryonic cells are determined. During this stage, some cells are set aside from the rest of the embryo to become the **germ cells**—the precursors of the sperm or eggs. The rest of the cells that will form the embryo begin to interact with one another to develop along three major cell routes. These lineages make up three layers that will develop into the body's different tissue and organ systems (figure 5.2):

- The **ectoderm** is the outermost layer of the embryo. It generates the surface layer of the skin (the epidermis) and also forms the brain and nervous system.
- The **endoderm** is the innermost layer of the embryo. It gives rise to the lining of the digestive tube and its associated organs (including the lungs).
- The **mesoderm** is sandwiched between the ectoderm and endoderm. It generates the blood, heart, kidneys, gonads, bones, muscles, and connective tissues (i.e., ligaments and cartilage).

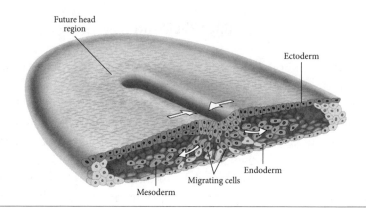

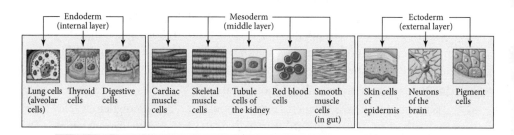

FIGURE 5.2 Gastrulation and the Formation of the Three Germ Layers.

The outer layer, the ectoderm, becomes the epidermis and nervous system; while the internal layer, the endoderm, generates the digestive tube, its accessory organs (liver, pancreas, etc.), and the lungs. Between them, the mesoderm, forms the circulatory system, the reproductive and urinary systems, as well as connective tissues and bones. The three layers interact with each other to form these organs.

ORGANOGENESIS

Once these three layers are established, their cells interact with one another and rearrange themselves into tissues and organs. This process is called **organogenesis**. Organs form rapidly during the first trimester (figure 1.1). The heart forms during week 4, and the first rudiments of the limbs (legs and arms) can be seen then, too. The eyes start to form during week 5, on the sides of the head; by week 7, the eyes are in the front of the face.

Organs form by cells interacting with each other. Organogenesis begins in the center of the embryo and works its way outward by progressive interactions among young cells. For instance, the mesoderm in the center of the back sends out chemicals to the ectodermal cells above it, and these chemicals tell the ectoderm, "You're not becoming epidermal skin; rather, you're going to become the neural tube and develop into the brain and spinal cord." The neural tube cells form, and the epidermal cells grow over them. Together, the neural tube and the mesoderm tell the cells on either side of them to become the vertebral bones, ribs, and back muscles. These cells interact with their neighbors and so on, forming the organs of the body. In the neural tube of the head, two groups of cells, one on each side, extend from what will become the brain. These cells touch the facial ectoderm and tell these cells, "You're not going to become facial skin; you'll become the lenses of the eyes." As these two lenses form, they tell the neural cells that formed them, "And you're not going to become brain cells; you're going to become the retinas." Thus, the eyes and other organs form by a process called **reciprocal induction**. The parts of each organ help the other parts to form. A major area of developmental biology investigates the chemicals involved in these body-forming dialogues.

Also during organogenesis, certain cells undergo long migrations from their place of origin to their final location. For instance, the precursors of eggs and sperm must migrate from the base of the yolk sac into the developing gonads—quite a long trip. The cells that form the facial bones migrate from the back of the head and neck.

By the end of the first trimester, all the major anatomical parts—including hands, feet, ears, and toes—are present, although many of them are not yet complete. All the organs except one have a fixed fate. This one exception is the gonad. Although the embryo's sex has been specified by its genotype (see box figure 3.1C), its sex organs are still rudimentary and "bipotential." The gonadal rudiment has the unique ability to become one of two organ types: the testes or ovaries. If the Y chromosome is present, a particular gene on that chromosome tells the bipotential gonad to begin its journey to become testis, not ovary. If there are two X chromosomes present (and no Y), the bipotential gonad starts the reactions that usually cause it to become an ovary (not testis). Around week 11, the germ cells migrate into the gonads and are told by the gonad to become either sperm cells or egg cells.

At the end of first trimester, the embryo is about four inches long and weighs about one ounce. By the end of the second trimester, the fetus is about twelve inches long and weighs about two pounds. What is really growing quickly now is the nervous system. The human electroencephalogram (EEG) pattern, a marker of brain functioning, is first detectable at around week 25 (i.e., around the beginning of the seventh month of pregnancy). Moreover, the human brain keeps growing at this rapid rate for about two more years after birth! This makes it different from other animals' brains. The brain of an adult chimpanzee doesn't have a brain much different from the one it was born with, whereas most of the human's brain neurons and the connections between brain neurons were formed after birth (Purves and Lichtman 1985; Greene and Copp 2014).

BIRTH

The human baby is born as soon as its last critical organ system—the lungs—matures. If development were to go on too long inside the mother's body, the baby's head would grow bigger than the birth canal, and it would not naturally be able to get out of the mother. However, if the fetus is born *before* the lungs mature, the baby will not be able to breathe on its own. The lungs start maturing at around week 25, about the same time as the human EEG pattern acquisition. To coordinate the timing of birth with fetal lung development, humans have evolved an intricate system whereby the mature lungs of the fetus send a signal to the uterus, telling the uterus that it's time to start contractions.

Like fertilization, birth is a process, and it is called (for good reason) **labor**. During the first stage of labor, contractions pull the cervix open and push the baby forward. In the second stage of labor, the baby is born. The third stage of labor pushes out the placenta (now considered the "afterbirth"), and the fourth stage is the recovery period.

When a baby takes its first breath, it changes the internal anatomy of the baby. The air pressure closes a flap in the heart, and this flap now separates the blood circulation to the lungs from the blood circulation to the rest of the body. The baby can now breathe on its own, and it can oxygenate its own blood. The umbilical cord, which had been the source of oxygen for the growing fetus, can now be cut.

Although many mammals are born functionally complete, the human infant is born very immature. Unlike horses, for instance, it cannot walk, and a baby cannot even find its mother's breasts without help. Even its eyes function only poorly. Many biologists think that this condition has been brought about as an evolutionary compromise between the growing head and the size of a woman's pelvis. As mentioned, the human brain keeps growing throughout childhood, making millions of new nerve cells each day. If humans were born at the same stage of brain development as their ape relatives, a baby would probably be born at around eighteen months, and its head would be far too large to pass through the birth canal. So it could be said that we spend the first few years of our lives as "extrauterine fetuses," totally dependent on parental care. We are kind of like kangaroos, only our pouches are cloth. We are still forming our nervous system during this time, and our ability to think and interact with others may arise from the fact that we are being socialized during a time when we are rapidly forming neurons and can learn extremely rapidly (Rose 1998; Gould 1977; Montagu 1962).

CONGENITAL ANOMALIES

Not all babies are born "perfect." Indeed, it is estimated that 5 percent of normal deliveries are of babies with a noticeable congenital anomaly, or birth defect. These can be lethal, such as defects in heart valves, or they can be benign, such as webbing between the toes. But with so many chemical and cellular interactions going on, it is not surprising that some things go wrong. It is often said among embryologists that the amazing thing is not that some babies are born with congenital anomalies. The amazing thing is that *everyone* isn't.

There are three main causes of congenital anomalies: bad genes, bad environment, and bad luck. Bad genes are mutations in the DNA that produce proteins that do not work properly. For instance, a mutation in the gene encoding the Tbx5 protein produces defects in the heart and in the fingers, since this protein is used to create the digit bones and the septum separating the heart ventricles.

"Bad environment" means that the fetus was exposed to chemicals that interrupted normal development. For instance, the drug thalidomide stops the development of arms, legs, and ears, and alcohol (even small amounts in some people) can disrupt the development of the nervous system and leave the baby cognitively impaired. Other compounds are now also thought to be responsible for birth defects. The Zika virus appears to infect the developing brain cells of fetuses, leaving them with smaller and less functional brains (Li et al. 2016; Tang et al. 2016). And endocrine disruptors may cause sterility. Bisphenol A (BPA), for instance, is a substance that is found in many of the plastics used to package our food and beverages. It has been shown to cause chromosome abnormalities in the eggs of mice, and people with higher than normal amounts of BPA in their blood are prone to miscarriages and failed pregnancies (Chen et al. 2013; Lathi et al. 2014). Interestingly, it has even been suggested that religious groups and scientists could form alliances to get rid of such toxic chemicals in our environment (Gilbert 2013).

Bad luck is another reason for congenital anomalies. People can have good genes, and good environments, but bad luck. The number of molecules made in the developing body is not constant. Some days more of a particular protein is made, some days less. There is a lot of randomness occurring in development, and if the embryo is making a low amount of a particular protein during a time when large amounts are needed, the cells might not migrate or differentiate correctly.

WHEN DOES AN INDIVIDUAL HUMAN LIFE BEGIN?

Scientific Discussions

One thing we can say with absolute certainty is that there is no agreement among scientists as to when an embryo becomes a person.

There are at least four stages of development that scientists have claimed as the point at which human life begins,[2] including the following:

1. Fertilization (the acquisition of a novel genome)
2. Gastrulation (when the embryo commits to becoming a single organism, and twinning is no longer possible)

3. EEG activation (the acquisition of the human-specific electroencephalo-gram, or brainwave, pattern)
4. The time of or surrounding birth (the acquisition of independent breathing and viability outside the mother's body)

View 1: You Become Human at Fertilization

In this "genetic" view of human life, a new individual is created at fertilization (conception), when the genes from two parents combine to form a new genome with unique properties. This perspective is the current official position of the Catholic Church, and it is the view held by many anti-abortion activists. For instance, several anti-abortion websites tell their audiences that every attribute of our life—intelligence, attractiveness, outgoingness—is determined by the genes we get when sperm meets egg. As previously discussed, however, that is not the case. Our intelligence and personalities are also formed by how we interact with our parents, who we meet as friends, our economic circumstances, our diet, and early love and traumas.

Many debates have focused on the issue of whether having a human genome is in fact the *sine qua non* of being a person. Anderson (2004) writes, "Does not the embryo possess all the genetic stuff of full humanness?" Sandel (2004) replies to this argument, writing that "the same thing can be said of a skin cell. And yet no one would argue that a skin cell is a person or that destroying it is tantamount to murder." Not all scientists believe that fertilization is when a person becomes human.

View 2: You Become Human at Gastrulation

American presidential contender and former Arkansas governor Michael Huckabee (2012) claimed, "Biologically, life begins at conception. That's irrefutable from a biological standpoint." However, this is merely another "spell" being cast on us that has no scientific validity. In reality, there are several other positions that scientists have taken. For instance, some scientists say that an individual human life begins at gastrulation. As mentioned, gastrulation is the process in which the embryonic cells begin the process of differentiation into the specific cell types of the new body. At this point, the embryo can no longer regulate to form identical twins. Because gastrulation is the point at which an embryo can give rise to only one person, many scientists consider gastrulation to be the stage at which an embryo becomes an individual. This is also around the time when the embryo becomes fully implanted into the uterus and initiates pregnancy.

This embryologic view is expressed by scientists such as Renfree (1982) and Grobstein (1988) and has been endorsed theologically by Ford (1988) and McCormick (1991), among others. Shannon and Wolter (1990) also raise the theological issue that, whatever ensoulment may be, it cannot happen before day 14, since, with twins, each is a distinctly different individual.

The view that a human does not become an individual before gastrulation, around day 14, is particularly crucial in the debate about allowing research on human embryonic stem cells, which we will discuss later. The embryologic view is consistent with the use of embryonic stem cells in biomedical research and has been supported as such by the conclusions of national commissions in numerous countries (including the Warnock Committee of the United Kingdom and the National Institutes of Health Human Embryo Research Panel in the United States; see Hyun et al [2016]). According to these committees, it is only at gastrulation that there is evidence of biological individuality.

View 3: You Become Human When You Acquire the Human EEG Pattern

This "neurological" view of human life looks for symmetry between the ways we define human life and human death. Think of all those movies where the doctors and family are standing around the patient's bed, looking at the electroencephalogram (EEG) tracings. Eventually, the spikes become flat, and the doctor notes the time of death. Several countries (including the United States) have defined the end of human life as the loss of the cerebral EEG pattern: Death is determined by the "flatlining" of the EEG, even though the patient may have a heartbeat and be breathing. The "neurological" argument proposes that if the *loss* of the human EEG pattern determines the end of life, then its *acquisition* (which takes place at about twenty-four to twenty-seven weeks) should be defined as the point at which a human life begins (Flower 1985; Morowitz and Trefil 1992).

The EEG pattern indicates that cerebral neurons are linked by *synaptic connections* in a manner characteristic of conscious brain activity. Just as a pile of unconnected microchips cannot function as a computer, the unconnected neurons of the fetal brain lack the capacity for conscious function prior to week 24. If one considers the quality of conscious awareness to define a human individual, this is a legitimate view of the starting point of a person's life. A human corpse is treated with respect and is different than the dead remains of other animals (one cannot, for instance, eat it). It is human, but it is no longer a person. It cannot vote or inherit. Indeed, as the writing in this paragraph demonstrates, the corpse is an "it" not a "who." Similarly, say the proponents of the EEG pattern as being critical for personhood, the human zygote, embryo, and second-trimester fetus are human, but they not yet persons.

View 4: You Become Human at or Near Birth

There is also the view that a fetus should be considered human when it can survive on its own. These scientists believe human life begins when an individual has become fully independent of the mother, with its own functioning circulatory, alimentary, nervous, and respiratory systems. This traditional "birthday" is often recognized by seeing the head of the baby emerge or having the umbilical cord cut.

Traditionally, the natural limit of such viability was imposed by the respiratory system—a fetus could not survive outside the womb until its lungs were sufficiently mature, which occurs at about twenty-eight weeks. Today, however, technological advances can enable an infant born as prematurely as twenty-four weeks to survive, although such infants are at high risk for having physical and/or mental disabilities.

One advantage of such moments is that they are well defined, public, and obvious: the crowning of the head, the cutting of the umbilical cord, the first breath, or the first cry. In the absence of a clear consensus on when life begins, there are people who feel that birth is the only indisputable moment at which an embryo becomes a person.

As mentioned, biomedical research indicates that (even without including induced abortions) less than 50 percent of human embryos conceived do not survive to birth. Of these, most of these embryos miscarry prior to the eighth week of pregnancy, and there is no assurance that any given egg, embryo, or fetus will survive to be born (Mantzouratou and Delhanty 2011). One professor (Sandel 2005) testified to the President's Council on Bioethics that our society is not prepared to value a fetus as person: "If the embryo loss that accompanies natural procreation were the moral equivalent of infant death, then pregnancy would have to be regarded as a public health crisis of epidemic proportions: Alleviating natural embryo loss would be a more urgent moral cause than abortion, in vitro fertilization, and stem cell research combined."

Thus, given that most embryos do not survive to be born, this group of scientists sees birth as the time when one becomes a person.

The Scientific Conclusions

Even with the immense knowledge of developmental biology now available, there is no consensus among scientists as to when human life begins. The stages of fertilization, gastrulation, brainwave acquisition, and independent viability each has its supporters. So, also, does the view that there is no point at which one can say an embryo has suddenly become human and that the whole question of when human life begins is framed from the religious perspective of "ensoulment" and thus cannot be answered scientifically. As geneticist Theodosius Dobzhansky (1976) remarked,

> The wish felt by many people to pinpoint such a stage probably stems from the belief that a soul, conceived as preternatural entity, descends upon a formerly soulless living stuff, and suddenly transforms the latter into human estate. I hope that modern theologians can accept the idea that the transformation is not sudden, but gradual.

Any or all of these perspectives can be useful for contemplating what a human life is. Now, let's look at some religious traditions. There's no consensus here, either.

Religious Views as to When One Becomes a Person

There is also disagreement among religious doctrines about when an individual human life begins. There are many nuanced positions within each religion, and this section outlines the major tenets.

Traditional Jewish Views

The Jewish interpretation of when human life begins is extracted from three sources: the Torah, Talmudic law, and rabbinical writings. Modern Judaism is far from monolithic, however, and includes a number of denominations that interpret the classical texts differently.

While the Torah does not directly discuss the beginning of human life or voluntary abortion, it does condemn miscarriage that results from violence toward a woman by an unrelated man. Exodus 21:22–23 states that if a man injures a woman such that she survives but the fetus is lost, the perpetrator is to be penalized with a fine to compensate the family. If, however, the woman dies as a result of the violence, the attacker must "give life for life" and is to be executed, but no fine is incurred (Jacobovits 1973). This passage has traditionally been interpreted to mean that killing a fetus is not equivalent to the murder of a human being and that human life, therefore, does not begin during the fetal stage of development.

Talmudic law, based on interpretations of the Bible, is not settled on the matter, either. In one instance, the baby is awarded equal status to the mother's only when the head of a full-term baby appears at birth. At that point, the fetus can no longer be sacrificed to save the mother, which is the main reason for allowing an abortion under Jewish law. Thus, before this crowning of the head, the appearance of the broadest part of the face (usually to the eyes) from the birth canal, the fetus has no legal rights as a human being.

However, in another section of the Talmud, the fetus is said to be formed only after forty days. Before that "it is like water." This belief is actually similar to a view taken in medieval Catholicism and is derived from the position taken by Aristotle. Aristotle said that there were three types of soul: vegetative, animal, and rational. The vegetative soul is common to all living things and relates to the ability to grow and reproduce. The animal soul is specifically zoological, and it is the principle of movement.

The rational soul is acquired only by humans. In Aristotle's view, the soul is not something separate from the body, but something the body develops. According to Aristotle, a man develops his rational soul around day 40, which is when the eyes come together at the front of the body and the body takes on a more human appearance.[3]

Early Christian Views

The teachings of Jesus as articulated in the four gospels do not specifically address the question of when life begins (although much is said about being "born again"). Likewise, the apostle Paul, whose epistles, along with the gospels, are the foundation of Christian doctrine (i.e., the New Testament), has no definitive instruction on this point.

As Christianity gained converts in the classical world, there was increased friction between the rational, pragmatic philosophies of Greco–Roman culture on one hand and the spiritual doctrines of Christianity on the other. Tertullian (AD 197), one of the founders of Christian doctrine, denounced contraception and abortion, along with women and marriage. Women, "the gateway to the devil," were responsible for original sin and the death of Christ. Abortion was murder:

> Murder being once for all forbidden, we may not destroy even the fetus in the womb. . . . To hinder a birth is merely a speedier murder; nor does it matter whether you take away a life that is born, or destroy one that is coming to the birth. That is a man which is going to be one; you have the fruit already in its seed.

This may be one of the earliest clear statements of the premise that life begins at the moment of conception. Tertullian did, however, recognize the need for abortions when necessary to save the life of the mother (Buss 1967; Bonner 1985).

Positions of the Roman Catholic Church

During its history, the Roman Catholic Church has held varying positions on the beginning of human life. For most of the Church's history, its thinkers viewed immediate ensoulment at conception as impossible. The doctors of the Church—St. Augustine, St. Albertus Magnus, and St. Thomas Aquinas—each agreed that abortion was homicide only after forty days. This was the time when Aristotle had declared a fetus to be "formed." Around 1140, when the monk Gratian compiled the authoritative canon law, he concluded that "abortion was homicide only when the fetus was formed." Before the time of formation, the conceptus (the product of fertilization) was not considered to be a fully ensouled human. Catholic doctrine as expounded by St. Thomas Aquinas (1225–1274) also followed the Aristotelian interpretation that a male fetus became ensouled at forty days after conception, whereas the female fetus became ensouled at ninety days (Tribe 1990). Aquinas (c. 1260) believed terminating a pregnancy prior to that time was sinful—a particularly grave form of birth control—but was not murder. His teacher, Albertus Magnus (1249), the only canonized embryologist, wrote explicitly that the termination of pregnancy during the first month could not be considered murder, since the fetus was not yet formed.

There were also Catholic leaders who took exception to Aristotelian thinking. In 1588, Pope Sixtus V mandated that the penalty for abortion or contraception was excommunication from the Church; however, his successor, Pope Gregory IX, returned the Church to the view that abortion of an unformed embryo was not homicide. And in 1758, fear for the souls of those embryos that might die in the uterus caused Monsignor Francesco Cangiamila (1758) to publish *Embryologia Sacra*. This book advocated in utero baptism using a syringe—a practice that probably led to more than a few accidental abortions.

However, the Aristotelian view of ensoulment remained by and large the official view of the Roman Catholic Church until the mid-nineteenth century, when Pope Pius IX (1869) again declared the punishment for abortion at any embryonic stage to be excommunication. Much of the support for his view was based on the idea that, since we cannot know with certainty the time at which human life begins, it should have protection from the earliest possible time, that of conception. Although it might not be ensouled, the fetus "is directed to the forming of men. Therefore, its ejection is anticipated homicide." More recent Catholic theologians have argued that the rational human soul is, in fact, in place at the time of conception, because such an infusion must be a divine act. This argument has much earlier roots, having been put forth in 1620 by physician Thomas Fienus, who claimed that the soul must be present immediately after conception in order to organize the material of the body (DeMarco 1984).

Today, Roman Catholic doctrine maintains the belief that animation or ensoulment is concurrent with the moment of conception. It also departs from the views of Tertullian and Augustine, who accepted the use of abortion when the mother's life was threatened. The modern Church asserts that "two deaths are better than one murder." The Instruction *Donum Vitae* (CDF 1987) specifically states that "the human being is to be respected and treated as a person from the moment of conception; and therefore from this same moment his rights as a person must be recognized, among which in the first place is the inviolable right of every innocent being to life." The Church has some problems with the notion of identical twins and whether they share the same soul.[4]

Some Protestant Viewpoints

The many Protestant denominations of Christianity have taken widely divergent stands on issues such as slavery, homosexuality, and the admission of women into the clergy, so it is not surprising that there would be wide differences of opinion between and within Protestant congregations as to when human life begins. The Evangelical Lutheran Church in the United States is very open about these differences, acknowledging the wide diversity of viewpoints among its members. While recognizing that holding different views can be dangerous to the Church community, the Lutheran Church sees informed conversation about these issues as being beneficial, holding the possibility of clarifying one's beliefs concerning the roles of family and children, and concerning individual freedom and its limitations (ELCA 1991).

The Presbyterian Church of the United States (PMA 1992) accepts abortion as a last resort. Their stand appears more concerned with reforming the social environment than with worrying about when human life begins: "The Christian community must be concerned about and address the circumstances that bring a woman to consider abortion as the best available option. Poverty, unjust societal realities, sexism, racism, and inadequate supportive relationships may render a woman virtually powerless to choose freely." It continues that the Presbyterian Church does not advocate abortion, but rather "acknowledges circumstances in a sinful world that may make abortion the least objectionable of difficult options."

Some Protestant denominations claim authoritative knowledge of what the Bible dictates and will attempt to change laws in accordance with their beliefs. Thus, Resolution Number 7, "On Human Embryonic and Stem Cell Research," adopted at the Southern Baptist Convention (1999) states that "The Bible teaches that human beings are made in the image and likeness of God (Genesis 1:27, 9:6), and protectable human life begins at fertilization."

As these three examples demonstrate, Protestant Churches span the entire spectrum of positions on the beginnings of human life.

Islamic Views

Islam also has no official consensus on when human life begins and when abortion is permitted. Like the early Christian position, based on Aristotle, Islamic law espouses a view that strictly forbids abortion after the embryo has acquired a soul, something said to take place any time between 40 and 120 days after conception (Tribe 1990). A recent survey of abortion laws in Muslim countries (Shapiro 2014) confirmed this heterogeneity. Of forty-seven surveyed countries, eighteen forbade abortion except at times when the woman's life was endangered by the pregnancy, while ten countries allowed abortion on request. In 1964, the Grand Mufti of Jordan declared that it was permissible to seek an abortion as long as the embryo was "unformed," which in his opinion was within 120 days of conception. In 1999, another major Muslim religious leader permitted Muslim women who were raped in Kosovo to take drugs that caused abortion. Several Islamic scholars contend that even then, abortion should be allowed only with the father's or husband's approval and only when a woman's life is endangered or if the woman has been raped (Buss 1967).

Eastern Religious Views

Hinduism, as practiced by millions in India, is a religion whose foundations are entrenched in the principle of ahimsa, or nonviolence. The practice of nonviolence is intrinsic to the Hindu belief in reincarnation—the repeated re-embodiment of the soul in different individuals and even different species. The karma (net cause-and-effect of one's choices and actions) generated in one's present life determines whether one's soul ascends to a higher level or descends to a lower state in its next existence. The ultimate goal is to attain a state of bliss and enlightenment such that the soul is released from the cycle of earthly reincarnations and becomes one with Brahma, the Creator.

Hinduism teaches that abortion at any point is an act of violence, resulting in bad karma that will thwart the soul's progress toward enlightenment. Throughout the Vedas (the classical Hindu religious texts), pejorative references to abortion abound; it is referred to variously as "womb murder" and "the murder of an unborn soul" (Tribe 1990).

However, the advent of ultrasound has changed things enormously, and the cultural and economic preference for sons has led to the selective abortion of thousands of female fetuses. Many Hindu leaders and women's rights advocates support bans against sex-selective abortions. (This issue is discussed more fully in the appendix.)

In Buddhism, the first of five precepts is to avoid killing or harming any living being. Because the philosophy diametrically opposes the destruction of any form of life, even abortion to save the life of the mother violates the Buddhist ideal of self-sacrifice (for the mother). Its price is believed to be the woman's entrapment in the perpetual cycle of birth and rebirth (Tribe 1990). The current Dalai Lama (1993) has castigated abortion as "negative" but said, "I think abortion should be approved or disapproved according to each circumstance."

Science and Con-science

When does individual human life begin? There is no consensus within science, and there are even multiple opinions within each religion. People will and should reach an answer that is meaningful for them, and most people indeed do so. However, in answering the question for themselves according to their own knowledge, experience, beliefs, and emotions, some people feel that they have also answered the question for everyone else. Such a mindset rejects the idea that there have always been diverse ways of thinking, that new information is constantly emerging, and that we are constantly reinterpreting our traditions as we integrate new knowledge and experiences into them.

One must also remember that the exercise of "conscience" can be both for or against abortion. While current debates have usually portrayed physicians refusing to perform abortions as acting from conscience, the fact remains that physicians providing abortion services also do so as acts of conscience. Physicians providing abortions most often do so because they are impelled by their consciences. They have seen women die from self-induced abortions and abortions done by unskilled providers, and they feel that abortion can be lifesaving to the woman and honors the dignity of humanity. They provide abortions even though they risk imprisonment, death threats, long court hearings, and social stigma (Joffe 1995; Harris 2012; Shane and Wilson 2013).

If nothing else, we have to approach this question with humility and a sense of wonder. Differences in opinion hold both promise and peril. As the Lutheran Church (ELCA 1991) has stated, "Our differences are deep and potentially divisive. However, they are also a gift that can lead us into constructive conversation about our faith and its implications for our life in the world."

6

TECHNOLOGICAL MOTHERHOOD

CLARA PINTO-CORREIA

Miracles are not contrary to Nature; but solely to what we know about Nature.

—St. Augustine, *City of God*

I n the past fifty years, we've invented several new ways of becoming a mother. However, we've hardly taken any time to join efforts in thinking about what we're doing to how our future is globally going to be shaped by our options. Therefore, while chapter 3 discussed the development of the embryo inside the pregnant woman, we should now concentrate here on the mother herself.

What do we mean by "mother" in this day and age? Certainly, for the sake of this volume, we'll be discussing the biological, not the social, mother. But even this concept has evolved many different perspectives, as the notion of motherhood has been acquiring an increasingly incredible complexity from the 1970s to the present. This brutal change ought to be puzzling to the human mind while assisted reproductive technologies (ART) keep making new modes of reproduction possible. Before the late twentieth century, the way babies were born had never changed at all. ART, on the other hand, keep proliferating and outdoing themselves, constantly perturbing everything reproduction ever stood for, raising a few people's hopes and confusing most people's minds. Keeping up with these changes remains a difficult challenge for anyone concerned with social issues.

SURROGACY

As with many others, the forthcoming tale featuring desperate infertile couples appears in the Bible. Genesis 16:1 and 16:2 state, "Abraham's wife, Sarah, had not given him children. But she owned an Egyptian slave called Hagar, and Sarah told Abraham, 'God did not allow me to have children. Go, then, see my servant. Maybe through her will I have a child.' And God heard Sarah's voice."

Had the same quest occurred some thousands of years later, Hagar would certainly receive a good lump sum for her complete surrogacy services. In December 1980, for the first time in the United States, a woman with the pseudonym Elizabeth Kane legally gave birth to a child created through artificial insemination destined for an infertile couple and received five thousand dollars (Kane 1998). By 1987, surrogate motherhood clinics from California to Alaska were offering their services, complete with doctors, lawyers, therapists, and insurance experts, for an average of thirty thousand dollars. What fraction of these payments was given to the women carrying the children was never truly disclosed.

Between 1999 and 2013, it is estimated that 18,400 infants were born using surrogates (often called "gestational carriers" on government websites; CDC 2016). We don't know how fast the technique spread to other countries, because not all nations make their infertility statistics and tribulations public. What we know for sure is that the use of surrogates soon became so popular worldwide that, by the 1990s, a Brazilian soap opera shown in many other places around the world was titled *Rental Belly*. People got so used to the term and the concept that it didn't come as much of a surprise to anyone when, in 2015, it was surmised that Cristianinho, the five-year-old mystery son of famous Portuguese soccer star Cristiano Ronaldo (McConnell 2010) had been gestating in a Brazilian rental belly. There are obvious dangers in learning your science through soap operas, but, in turn, instant-hit soap operas like this one don't get their twisted plots from thin air.

Some of the First Worries

To understand why surrogates are still forbidden in a good number of developed countries, we have to remember at least a bit of what was initially unpleasant in their history. Surrogacy has been around even longer than in vitro fertilization (IVF), since surrogacy benefited directly from the artificial regulation of the hormone cycle developed to implement IVF but was easier to do. Surrogacy was still relatively rare when doctors began questioning what should be done, for instance, in cases of pregnancy complications developing with a surrogate. Should this brand-new third party be forced to undergo amniocentesis to monitor fetal abnormalities? And then what? Could the couple require the surrogate to undergo an abortion if there was unequivocal evidence of fetal deformity? Or what if a malformed child were born? Sure enough, speculations rapidly became reality. In 1977, a Michigan couple arranged a surrogate contract with a seemingly responsible Tennessee woman who turned out to be an alcoholic. In 1978, she gave birth to a child with fetal alcohol syndrome, an alcohol-induced birth defect with a prognosis of impaired intellectual development. And, in January 1983, a surrogate mother in Michigan gave birth to a child with hydrocephaly and a prognosis of severe intellectual disability. It is unknown how the parties involved settled these issues—and, even if all had agreed, it is possible that courts of law could have been of different opinion (Volpe 1987).

It could get worse, and it did. In 1985, an infertile couple paid over eight thousand dollars for another woman to carry their child since the wife was unable to conceive, although her eggs appeared to be perfectly viable. When the baby girl was born, her first blood tests promptly revealed that she was not the daughter of the childless husband, but rather of the surrogate's husband. But the couple decided to keep this other couple's biological baby anyway.

The gates had been swung open, and no one knew what was going to come out. In 1986, taking everybody by surprise, the press avidly covered a suddenly hot topic from what was then West Germany: a prostitute who had paid to gestate a baby for a rich couple using her own eggs and the husband's sperm didn't want to surrender the baby when he was born, claiming that he had come from inside her body and was therefore her child. She later won her case in court, since the law clearly stipulated that, when in doubt, the mother is the woman who delivers. Widespread pictures of the new mom in the hospital holding the newborn in her arms with her face heavily covered in makeup settled deeply in the public's imagination, and more stories of prostitutes recruited as surrogates circulated, while rumors grew that there was already a slave trade of "rental bellies" coming from the Philippines, although this was never possible to confirm.

With the lucid eye of the insider, Volpe (1987) summarized better than most these early agonies of the closing millennium:

> Surrogacy is technologically feasible, but is it morally defensible? . . . The carrying of a child is traditionally an intimate affair, and it should be an affront to one's moral and aesthetic sensibility if such an event is blatantly commercialized. . . . Scientists themselves acknowledge that surrogacy is not a new medical advance but rather a commercial enterprise wherein motherhood is determined by contract. The only "new" development is the hiring of attorneys who, for a fee, recruit women who are willing, for a price, to allow themselves to be used as human incubators and are willing to relinquish their gestational rights to the child.

The association of prostitution with surrogacy, based on the argument that there is no difference between selling sex and selling reproduction, was still heatedly debated by ethicists, feminists, doctors, and many others by 1995, and the fact that it is hardly possible to come to a general agreement on the subject remains behind the ban that several countries still impose on the procedure (van Niekerk and van Zyl 1995). Of course, legislation has evolved since the early days, and so has all the legal paperwork that surrogates have to sign in order to register at specialized clinics—but since profit has always been openly the name of the game, the claim that "motherhood is becoming a new branch of female prostitution with the help of scientists who want access to the womb for experimentation and power" (Dworkin 1983) was not uncommon at all at first. It seemed supported from the very beginning by the still oft-mentioned fact that a good number of women enrolled in surrogacy, regardless of how much profit they would make, ended up getting caught up in the strong connection with the baby developed during pregnancy, and even more so right after birth (Lorenceau et al. 2014; Papaligoura et al. 2015).

Then there was Baby M, whose story seemed made to order to illustrate these dramas. Mary Beth Whitehead, a surrogate mother, refused to give up the child she was contracted to bear, using both her egg and her womb (Whitehead and Schwartz-Nobel 1989). Some people advocated for Whitehead, giving supremacy to the maternal instinct created by motherhood. Others considered Whitehead untrustworthy since she had broken her contract, favoring the wealthy couple who had hired her. Many said the husband's contribution was the defining factor in the matter. Finally, it became known that Elizabeth Stern, the childless woman who had hired the surrogate, was not infertile but rather afflicted by multiple sclerosis, a condition likely to be worsened by pregnancy—and this was enough to silence most commentators. The narrative had gotten very complicated. It seems that no one rule serves all conditions. We may desire a uniform picture, but what nature gives us instead are interesting, complex, and ever-exceptional cases.

Modern-Day Trends in Western Surrogacy

Several countries have since legalized surrogacy, and some, such as the United Kingdom, have even developed entire welfare systems meant to ensure the well-being of surrogates and that of their children. Apparently, thanks to all such measures, surrogates feel proud of the happiness they've bestowed upon others, and their own children are generally proud of their mothers for having helped their fellow human beings in this most generous way.

However, even this rosy picture comes with another, much sadder side. Ever since the Internet began to provide surrogates with their largest mutual support website (www.surromomonline.com), it has become possible to uncover some of the thoughts and feelings of surrogate mothers. For instance, it is devastating to a surrogate if, for any reason, the pregnancy fails to produce a healthy baby. We tend to think only about the tribulations of the couple who hired the surrogate. But such loss can also be horrible for the surrogate herself.

Also, as in any other IVF patient, suffering with pregnancy complications and loss seems to be heightened owing to the increased expectations of success raised by all the attempts to maximize results (such as the transfer of multiple fertilized eggs and the early monitoring and testing for pregnancy after embryo transfer) (Berend 2010). But, being "only" surrogates, why should these women care? Because, by now, this notion of the gift of life, of being the vehicle for somebody else's fulfillment and happiness, of filling the empty cradle of a couple who has long been in silent mourning, has become a deeply rooted fantasy, at times obsessive to the point of neurosis. Reading through the surrogates' poignant statements online, we are taken aback by the degree of pain and sorrow at each loss of an embryo that these women were not even going to gestate for themselves. Apparently, there is no one out there to give them the support they need. Couples undergoing infertility treatments often have a therapist by their side. Do surrogates? Much still needs to be done if surrogacy is ever going to become a healthy part of any social fabric. In the United States alone, 30 percent or more of gestational carriers are poor black women servicing well-to-do white couples (Wilson 2014). The couples often have many more resources available to them than the women gestating their embryos.

What the Mother and Fetus Tell Each Other

In all these cases, it seems as though most couples think only about their genes and totally overlook the interactions that the fetus carrying those genes is necessarily going to have with the hidden world around it. This is the surrogate's world: what she eats, what she drinks, what she smokes, what makes her happy, and certainly what causes her any kind of stress.

Surrogate motherhood has been around for many decades now, and it has been the source of much gossip and diatribe (Steiner 2013). Yet, it is somehow still mainly regarded as simply an additional step in a biomedical protocol, not much different than a period of cell growth within a hospital incubator. But to hold this view is to grossly overlook what we also all know, at least at some level: that the prenatal environment has a major impact on the development of a baby and that the genes are not the whole story. Obvious questions immediately come to mind. For instance, we all know that pregnant women are not supposed to smoke, drink, do drugs, or even take a great number of over-the-counter medications. Doing any of these things could severely affect the baby by affecting the fetus' developmental environment. Now, what if a surrogate goes ahead and does all of the above? And what if her diet is profoundly unhealthy? What if for some reason she ends up in jail or becomes traumatically stressed? What if her companion beats her? Sure enough, all of these things can affect the future individual still unborn. But how are we going to prevent such misfortunes from occurring? Are surrogate mothers to be kept under constant scrutiny, or even locked up in a convent of recluse nuns to make sure nothing disastrous happens to them?

There is a new field of biology called ecological developmental biology, which looks at the relationship between the developing organism and its environment (Gilbert and Epel 2015). One aspect of this new discipline concerns the developmental origins of adult-onset disease. And yes, it really does seem that genes aren't the only agents that give adults their characteristics. The maternal environment plays a large role in this lottery. In the 1980s, David Barker and others provided epidemiological evidence that the number of calories a woman consumes during pregnancy can affect the metabolism of the fetus (Barker 1994). Not only that, these alterations in metabolism continue in the adult. Babies who are fed poorly in the uterus tend to be born with lower than normal birth weights and have a much greater risk of having heart attacks or strokes as adults. This phenomenon has been confirmed by observations in mice and rats. Moreover, the molecular basis for this phenomenon has been found. A low-calorie diet instructs the fetus that it will probably be born into a low-calorie environment (Gilbert and Epel 2015; Burdge et al. 2006). Therefore, the genes encoding the enzymes for rapid utilization of nutrients are suppressed, and the genes encoding enzymes that store nutrients are activated. From the evolutionary point of view, these mechanisms make perfect sense: they work wonders for a baby who is indeed born into a food-poor environment. But, in mice, if the fetus develops into a mouse born into a well-stocked environment, the storage genes are still activated. As a result, the mouse becomes obese and prone to diabetes and heart attack. In humans, this has been called the "mismatch hypothesis" (Gluckman and Hanson 2007).

So, there is reason for concern about how the surrogate treats the fetus. It is quite predictable that the child will have the eyes, skin color, and facial structure of the parents who provided the embryo. But the baby will also have all sorts of messages imprinted upon the fetus by the surrogate during pregnancy, and these can affect personality and health. Those messages will show up during that child's entire life. Genes aren't everything. Their expression works up against the bittersweet beauty of evolution at work. Get ready for more about this in the last chapters of this book.

EGG DONATION

Surrogacy initially involved making children using another woman's belly. In the early 1980s, we acquired the ability of making babies using another woman's eggs. And the woman supplying the eggs did not have to be the surrogate mother or the woman of the couple paying her. The possibility of using young and fertile egg donors appeared in 1983 and was meant to help women who had lost their eggs for chemical or natural reasons. These women were unable to ovulate by the time they first entered the infertility clinic, owing to their age, having undergone cancer treatments or pelvic surgeries, or other complications. Could they obtain donors to give them their eggs?

It was quite self-evident from the start that the term "donor" was little more than a euphemism. Whether or not these now-widespread "donors" really "donate," or rather *sell*, their eggs is generally not even mentioned, just as payments for "donations" tend to be obscured by the clinicians offering these procedures.[1] As for the concept of the "anonymous donor," it didn't take a long time to make its way from anonymity to colorful catalogues with all sorts of indications concerning the "donor" place and date of birth, hobbies, studies, interests, and the like—exactly as had previously happened with sperm "donation." Once again, lawyers, ethicists, theologians, and scientists worldwide remained hopelessly incapable of reaching any kind of agreement on how such matters should (or should not) be regulated. Research from 1999 already mentioned, by name, an American clinic where each "donor" was paid $2,500 per superovulation cycle (Angier 1999a, 8)—an extremely meager percentage of what a couple would have to pay for IVF with egg donation around that time. As usual, the "donors" at this clinic were all young women with strong fertility records having passed all required tests to prove themselves disease-free.[2]

POSTMENOPAUSAL PREGNANCY

Most people agree that something was off the mark in 2006, when the news included a story (Goldenberg 2006) about a postmenopausal woman, diabetic and blind from birth, already the mother of eleven, grandmother of twenty, and great-grandmother of three, who had just given birth to twins at age sixty-two with the help of another woman's egg, her husband's sperm, and IVF. Stated this way, menopausal motherhood certainly seems crazy. But, outside specialized literature, it is rare to find anyone describing the concept of having children after forty as a source of social concern rather than a triumph of medicine, or of the many wars fought in the ongoing battle for women's emancipation.

A Complex Story Made Too Easy

These latest biomedical possibilities are certainly seducing, and honestly, let he who is without sin cast the first stone. They understandably often come across as something of a fantasy to anyone who raised small children while working a demanding job all day long. Just think of the increasing number of famous women past their forties proudly posing in glossy magazines, and even on TV ads, with their newborn babies and their promptly recovered elegant silhouettes. By the twenty-first century, the possibilities offered by several ART combined eventually led to the ongoing trend of "yummy mummies," women who desired both to keep their good looks and to have a child with one of their own eggs. This trend has been on the rise since the late 1990s, without it being all that clear to the average person how one could achieve late motherhood so easily. There was Julia Roberts with twins at thirty-eight, Jane Seymour with twins at forty-four, Susan Sarandon with a baby at forty-six, Geena Davis with twins at forty-seven, and Holly Hunter with twins at forty-seven. By then, the American public had already come to think this all normal.

And what else could people think? This was entertainment news, not science reporting. Nobody expected the new forty-something moms to be featured alongside the description of the techniques used to bring their babies about. These famous moms were rather, and quite simply, perceived as the pioneers of a new way to ensure professional success first and then proceed to the long-postponed joys of having babies, hence finally solving the infamous work-versus-family dilemma that had become a hallmark of the last century.

Not all celebrities remained silent on the methods they used, but their explanations, for all of their certainly good intentions, did not always clarify to the public how they did it. In 2003, while she was the anchor of *Good Morning America*, Joan Lunden set a complicated precedent when she went public about having used a surrogate at age 52. She had first raised a family during the typical fertile years but then remarried a younger man and decided to do it all over again. She first "had" twins at fifty-two. Two years later, she "had" twins again. She enthusiastically encouraged other women her age to do likewise. However—and this is where other women her age might get confused—she never revealed whether or not she had also used egg donors (although, at fifty, this option seems very likely, unless some awkward miracle happened twice.) Later on, as other famous women in their late forties such as Nicole Kidman and Sarah Jessica Parker started ranking among those who openly discussed their use of a surrogate to have children, the practice began to feel (or seem) increasingly familiar.

Medical Risks of Late Pregnancies

Obviously, when famous women under constant scrutiny choose to keep delicate details of their private lives as quiet as possible, they are not to be blamed for that choice. Privacy is a sacred right we all want to see respected when it comes to our own lives. The real problem resulting from this situation is what many women are led to believe from what they read about the lives of their powerful, perennially young-looking counterparts. These days, bombarded by all the happy baby stories force-fed to them on the news, even bright college students seem to assume that pregnancy after forty is easy to achieve. But exactly for the sake of all these bright-eyed young women willing to believe in love at first sight and marriage for life, and who are already considering how to plan their pregnancies in order to create the best family possible, it should be known that there are risks and shortcomings in leaving babies for later. Much as any Millennial might dream of organizing her material life first and only raising her kids afterward so that she can give them the best of all possible worlds, dreams are still dreams—and the German word for dream is still *Traum*. I would hate to see my beautiful young students traumatized by the failure of their best adult intentions. Hence the urgency of the situation.

Pregnancy after age forty is still not a doctor's idea of the best way to have a baby. By 2015, papers bearing such long titles as "Family Intentions and Personal Considerations on Postponing Childrearing in Childless Cohabiting and Single Women Aged 35 to 43 Seeking Fertility Assessment and Counseling" clearly reveal a trend toward delaying pregnancy. The titles also show that doctors know that achieving a successful pregnancy becomes more difficult the longer the pregnancy is postponed.

To start with, doctors know that such late pregnancies are a serious gamble against natural odds (Roy et al. 2014, 56): "As a woman ages her ovaries become less able to release eggs, she has a lower number of eggs left, these eggs are less healthy, and she is more likely to miscarry." There are also be additional health risks for the mother, such as diabetes, hypertension, and womb-related cancers.

One of late pregnancy's most substantial risks, though, is the occurrence of chromosomal disorders in the baby owing to the mother's age. Remember, the same oocytes have been in a woman's ovaries for over forty years. The cohesion proteins that hold the chromosomes together have not been renewed, and like any person's forty-year-old proteins, they can come apart. Most fetuses with chromosomal disorders die. Others endure sexual anomalies (including infertility). Down syndrome (resulting from an extra copy of chromosome 21) is one of the few chromosomal anomalies that allows a fetus to survive. This can occur in older women with such frequency that, in the early 2000s, some infertility clinics started refusing to perform IVF in women forty or older. Therefore, it is quite likely that those happy celebrities on magazine covers had legitimate reasons to keep a low profile, but, in doing so, they were not telling the whole story of how their babies were born—with a surrogate and from a donor egg. We may have come to profoundly dislike this idea, but we haven't truly beaten nature at its own game—certainly not yet.

OOCYTE CRYOPRESERVATION

Maybe we should phrase that last statement in a more significant, definitely more modern, way. At least where fertilization and embryonic development are concerned, until a couple of years ago, we *were* under the impression that we truly couldn't beat nature at its own game—or, at least, not yet. However, another new step challenging what used to be the norm for natural reproductive limitations was to be addressed next, once more in what seemed to be no time at all. And, once more, this seemingly well-meaning biomedical addition to the already copious ART catalogue ended up having its own way of becoming quite disruptive from the social and emotional point of view.

The recent success in *egg freezing*—actually "oocyte cryopreservation," but "egg freezing" sounds much less clinical—is indeed a great breakthrough, since we were previously able to freeze only sperm and embryos. Most likely owing to their much higher fragility, oocyte freezing, and later thawing, had remained impossible until 1972, and then the result was not repeated until the early 2000s. Initially, the process was aimed at three particular groups of women and was therefore designed as both a therapeutic and moral alternative to all other types of ART available on the market.

In the group of those initially benefitting from oocyte cryopreservation, first came women diagnosed with cancer who had not yet begun chemotherapy or radiotherapy. As mentioned, chemotherapy and radiotherapy are toxic for unfertilized eggs, leaving few, if any, viable cells behind. Every year, fifty thousand reproductive-age women are diagnosed with cancer in the United States alone. If you happened to belong to this group, wouldn't you like to avoid losing your reproductive material for good by resorting to oocyte harvesting followed by cryopreservation? Oocyte freezing offers women with cancer the chance to preserve their reproductive reserves, so that they can have children in the future by undergoing treatment with ART.

Second, there were women who would like to preserve their ability to have children for later, either because they did not yet have a suitable partner or for other personal or medical reasons. For instance, women with a family history of early menopause often have an obvious interest in fertility preservation. With oocyte freezing, they would have a store of this precious resource even if they underwent menopause at an early age.

And third, the technique was finally an acceptable choice for women and doctors alike who consider embryo freezing morally unacceptable. Just like sperm freezing, oocyte freezing does provide an option in which moral issues just about disappear. The cells frozen by this technique are by no means different from any of those hundreds of oocytes women shed monthly from menarche to menopause every time they menstruate. In any given menstrual cycle, oocytes die if they are not fertilized, and this obviously happens more often than not. Therefore, for all those people objecting to supernumerary frozen embryos, having the possibility to fertilize only as many oocytes as will be used in the IVF process and then freeze any of those remaining unused, can be a positive and nonthreatening solution from a bioethical point of view.

Egg freezing is an expensive technology and brings with it all the dangers of ovarian hyperstimulation, as well as pelvic and abdominal pain (see the appendix for more on these issues). So why would anyone not medically in need of this procedure want it?

The Modern Narrative of Yuppie "Frozen Eggs"

It turns out that, with time, a lot of young women seem to decide they needed this technique badly for reasons having nothing to do with their health, and that's where the fireworks started. There is quite a sensational part to this seemingly quiet story, but it erupted only in 2014. While most private insurance companies in the United States cover few or none of the expenses relating to infertility treatments and ART, both Facebook and Apple announced they were including oocyte freezing in their total benefits packages for their employees (Zoll 2014). The idea seemed to be exciting enough for the general public to be promptly told an explosive story that goes more or less like this: Upon signing your contract with these companies, you could order a hormone package and put your oocytes away in the freezer for whenever you were ready to use them—hopefully in your forties, so that during your energy-loaded, creativity-heightened twenties and thirties, you are not encumbered by those complex mood-swinging, time-consuming processes of pregnancy, delivery, and (if you dare) maternity leave—and then all the subsequent years filled with the troubles and tribulations of raising those ever-demanding children, who obviously distract you from your work and immediately twist your priorities. Or, in cruder terms, so that corporate America makes the most of you, and vice versa.

In a friendly move seemingly supporting this tale, the American Society for Reproductive Medicine (ASRM) removed the experimental label from the procedure by early 2013, so that anyone needing and willing to pay for oocyte freezing could go ahead and do it. What's interesting, though, is that this new permission was for oocyte freezing for people with medical reasons for undergoing the procedure. The ASRM stated: "There are not yet sufficient data to recommend oocyte cryopreservation [egg freezing] for the sole purpose of circumventing reproductive aging in healthy women because there are no data to support the safety, efficacy, ethics, emotional risks, and cost-effectiveness of oocyte cryopreservation for this indication" (Practice Committees of the ASRM and the Society for Assisted Reproductive Technology 2013). And in the same report, which lifted the ban on medical uses of this procedure, the ASRM wrote, "Marketing this technology for the purpose of deferring childbearing may give women false hope and encourage women to delay childbearing." Indeed, the data supporting egg freezing were from women aged thirty-five and younger.

Was anyone listening, though? There was certainly a positive financial response to the newest trend in ART, regardless of all the cautionary words coming from medical authorities. By the summer of 2015, the costs of oocyte-freezing clinics were already falling from the initial ten thousand dollars, especially after the announcements from Facebook and Apple.

And the news kept getting stranger. The story moved in a flash from the newspapers to blog after blog and caught fire on social networks, creating a truly bizarre picture of this new industry and its users. Information sessions began taking place in plush, trendy hotels that went by the name of "egg freezing parties." Women, we were told, could buy their youthful time for decades, in order to enjoy great Caribbean vacations with plenty of booze and guys now. While relaxing and partying, the story went, they were postponing the nuisance of screaming kids until much later. In the meantime, they could dispense with the weight of permanent partners and show up for work in all their glory to climb up the corporate ladder wearing smooth-operator suits. Soap operas were already taking note.

From the beginning of what might be the weirdest ART fairy tale ever, it was remarkable how this news circulated without it ever being acknowledged that no one knew how long a frozen human egg would retain its viability or fertilizability. And the data do not look good. Even in young, healthy women (younger than thirty-eight years old), the chance that a frozen egg will yield a baby in the future is estimated somewhere between 2 and 12 percent. According to the American Society for Reproductive Medicine (2014), one can expect a 90 percent failure rate. If this estimate is anywhere near correct, most egg-freezing clients will therefore be forced to deal with a future of childlessness whenever they finally choose to have children. How many oocyte-freezing customers had been clearly told about this serious shortcoming remains anyone's guess. It appears as though the oocyte cryopreservation priority of these last years has moved away from its original medical and religious reasons and is rather centered primarily on blatantly persuading young women that they can—and should—always have it both ways.

Egg Freezing Goes Mainstream

Most people believe in advertising, perhaps because persuasion comes with a welcomed soothing effect. In the case of egg freezing, different priorities clearly did make a social difference. When the first survey of the motivations of the technique's users was carried out in New York, only a minority of the women interviewed mentioned postponing childrearing for the sake of fun or their career as the main reason to freeze their oocytes. However, by 2014, after everybody read all the stories circulating online, which created a formidable wildfire, the technique seemed to have become increasingly trendy as a much-welcomed break for working girls. In a survey of more than 560 women younger than thirty-four published in *Cosmopolitan* in late 2014, over half said that, in addition to taking off their shoulders the immediate pressure to find a partner, they would consider oocyte freezing in order to have as much fun as possible before having kids, or so that a baby wouldn't derange their careers early in their lives. Also, the social ecosystem these young women belonged to had apparently taken this option for granted. In the spring of 2014, the *Bloomberg Business* cover featured the headline, "Later, Baby? Will Egg Freezing Free Your Career?" The number of women believing the answer is yes and stepping forward with the thousands and thousands of dollars these lotteries cost keeps increasing as time passes: their answer is yes, yes, yes. Abby Rabinowitz (2015) researched several clinics and found that the going rate for freezing eggs was around forty thousand dollars. As the experts say at think-tank meetings, "Advertising is the fine art of separating people from their money."

Maybe another way that these techniques have gone mainstream is by being featured in American sitcoms such as ABC's *Modern Family*. In an episode from 2015, there is a brief appearance by a female lawyer successful to the point of having photos of herself with Michelle Obama and with Maya Angelou on the walls of her office. She tells her new partner that, yes, professionally her life has been a success. However, she hasn't had a date in six months. Then she makes a dreamy face: "Sometimes, on Sundays, I take a ride to go visit my frozen eggs." What was this gorgeous creature wearing? As the stereotype mandates, an extra-sharp turquoise corporate suit complete with the unavoidable matching high heels. She certainly looked fabulous; yet she never appeared again. It makes you wonder, were viewers minimally aware of the full scope of implications at stake in this woman's choices—or could this possibly have been yet another tool of persuasion? And who's to tell us?

A Matter of Profit

It is not even possible to evaluate how much success egg freezing has truly registered so far, since figures keep changing in the overall estimates that different clinics claim for their services. Companies backing the laboratories involved candidly admit that the IVF market didn't have much more room to grow, and thus they had seized a new source of revenue in the oocyte-freezing business (Spar 2006; Rabinowitz 2015). And the target is equally obvious: young, hard-working, intelligent women with graduate degrees, mostly single, often looking for fun with a small gang of girl-friends, and generally highly concerned with their looks. We are now in the age of "egg freezing parties" (box 6.1).

BOX 6.1: EGG-FREEZING MEETS *SEX AND THE CITY*

In 2015, a smart potential client (Rabinowitz 2015) writing for *Nautilus* described one of her personal close encounters with the industry with memorable passages such as the following:

> Last fall, I went to an egg freezing cocktail hour. The downstairs bar of the glossy SoHo hotel was thronged with women in heels and sleek business attire. Club music thumped, cameras flashed, and I narrowly missed being hit by a videographer angling a tripod over the crowd. The evening was hosted by . . . a startup that sells financing for egg freezing, framed as fertility insurance for the forward-thinking urban professional woman.

Having herself already frozen oocytes once, she vowed never to do it again, because, as the head of a clinic had sagely told her, "An insurance policy guarantees to pay if your house burns down," and therefore "egg freezing is not an insurance policy; it's a lottery." However, the author later finds herself riding the subway home alone and contemplating the bag with shiny gadgets offered at the party: "I found myself contemplating egg freezing again, based on a biomedical marketing event dressed up as a girl's night out in *Sex and the City*."

The theme of one posh Manhattan hotel party was "Fun, Fertility, and Freeze" (Johnston and Zoll 2014; Lambert 2015) The ad for this party on the subway is "To Emma (Age 42). Love Emma (Age 30). If you are not ready to have a baby, freeze your eggs now and give yourself the gift of time." What exactly is going on in all these stories? Spells are being cast.

Passing judgement is easy. Understanding is much harder. It requires a personal and collective effort, but it's becoming increasingly necessary, almost a civic duty. The truth remains that the leading reproductive scientists in the United States estimate that most of the young women who freeze their oocytes for a later age will not be able to use them. This can be devastating. These women are being promised the miracle of living like their male counterparts for as long as they want to and then becoming doting mothers when the time is perfectly right to start a family. But if this doesn't happen, what then? Let's keep in mind that "a decrease in fertility begins at age thirty, and by forty years old the chances of getting pregnant dramatically drop" (Roy 2014, 56). If an aspiring forty-year-old woman can't get pregnant anymore, then it's again off to the surrogate or the egg donor. But then, as with any other pregnancy, we still have to verify whether the embryo grown in vitro will successfully nest in the womb or survive the first trimester—and, as we already mentioned, in real life, this is not likely to happen right away.

Some Collective Social Responsibilities

Having by now heard many times that Denmark continues to rank annually as the happiest country on the planet (it has been ranked first since 1973), some of us would consider it reasonable to attribute this happiness to the sense of security people get from belonging to a solid welfare state. Bring this up with the Danes, and they will agree. Now, a solid welfare state like theirs means one year of paid maternity leave for both parents, one full month of paid vacation time, a generous stipend per child until they leave the household, an extremely good education system that willingly takes your children in from a very early age—some of those important things that their huge taxes are meant for. It makes you almost assume that, with all this help, they don't need to postpone having children until a much later age. "Sorry," promptly corrects one of my Danish friends, "My daughter is thirty-eight and she still has no children. She has all those benefits, yes, but she works for a multinational corporation. The pressure on her to take work home and stay up late is enormous, not to mention all those business trips she has to take all the time. She's afraid of having children because of her job. Like everywhere else in Western-type societies, I guess."

All these different factors tell us that most techniques already in use will continue to be used, and there seems to be no reversing the tide. Regardless of whether or not we like the idea, the sooner we deal with the concepts of menopausal moms, surrogacy, egg donors, and egg freezing, the better. As a culture, we should at least ask ourselves where we are going and whether we should really go there. For instance, in my generation, many thirty-year-old single mothers working full-time like myself were already sniffing one white powder or another to keep up with the challenge and be happy and energetic enough to handle their kids' demands before and after their high-performance jobs. I could easily understand why. What now? Will menopausal moms somewhere in their late fifties have patience for their restless triplet toddlers when the time comes, not to mention the energy required somewhere in their late sixties to deal with the aches and pains of teenagehood—and are we willing to play it dumb and concede that they will all still be alive by then? Is medicine really going to immediately allow us all to remain alive, energetic, and young at heart far longer than ever—or are we ready to fill the world with orphans or with children of addicted mothers just because *we can*?

THE "BIOLOGICAL CHILD" DILEMMA

As we make our way through the expansion of ART, we still haven't addressed the most obvious question of them all, and it's about time we do it, much as we're likely facing the hardest step of our journey through the history of families. Still, by now the question begs itself. Just why on earth are people putting themselves through all this hardship, all these expenses, all these shortcomings, all these endless broken hopes, to get themselves what they stubbornly consider to be a genetically similar child? We have verified time and again that these ART babies are often going to differ from their parents in one way or another. If they used donor eggs or sperm, the child is not going to be that similar genetically. If they used a surrogate, the conditions of the womb aren't going to bear their developmental messages. There is a costly and somewhat odd fixation on biological legacies out there, and it is built around fantasies. By now, it has clearly become a genetic neurosis.

So why don't people avoid all these expensive and anxiety-producing procedures and, for instance, simply adopt a child and go on with their lives? American couples just aren't adopting children as they used to (Wilson 2014). Prospective parents often mention that adoption is risky, since the biological parents may decide to keep the child after it is born, and the mother, often from a low socioeconomic stratum, may be thought to be drinking or taking drugs. American couples also mention that the adoption procedure can be expensive, as adoption agencies charge for their services, and it can be very costly and time-consuming to bring in a child from overseas. It's almost as though the concept of who saves whom has been turned upside down. Adoption in America seems to have come to be perceived no longer as a child's salvation but as a couple's potential damnation, because of the bad genes or the bad past that could ruin the family. On the other hand, a biological child could be a couple's only possible salvation, radically turning the tables on what old social constructs used to be (Wilson 2014, 17).

It is commonplace that adoptees Bill Clinton, Eleanor Roosevelt, Gerald Ford, Willie Nelson, Nelson Mandela, and Steve Jobs did quite well for themselves. However, even the suggestion of adoption as an alternative, let alone a *reasonable* alternative, might be seen as unsolicited and opinionated. Certainly, for some, it remains an expensive option, but it is usually less expensive than IVF, especially after several rounds of hormones, which are typical in older women. Moreover, it seems only fair to point out that the increasingly complex combination of these techniques can raise a fair share of biological perplexity when people insist they are just trying to have "their own" child. In a world whose resources are being threatened by massive overpopulation, adoption seems both reasonable and moral. But we tend to view high tech as progress, the genome seems to be more important than parental attitudes in raising children, and there is not much profit incentive in adoption. The neurosis is as much a part of society as it is in our brains.

IV

IMPROVING THE HUMAN CONDITION THROUGH BIOLOGY

The Reality and the Fantasy

If one really wanted a genetically similar child, the way to go would be cloning. The result would be an offspring with the same nuclear genes as the parent. These chapters discuss the science of cloning and also the fantasy of cloning. Both Scott and Clara have worked in laboratories that have pioneered nuclear transplantation, and both have manipulated mammalian eggs to give them different nuclei.

Every technology has its history. Scott's chapter 7 looks at the technologies of animal cloning and how they changed when it became apparent that one could perform some of the medical tasks of cloning with embryonic stem cells. These stem cells were difficult to obtain (and morally worrisome to many). Moreover, a new technology—induced pluripotent stem cells—has enabled researchers to transform nearly any cell of the body into an embryonic stem cell. This has brought new worries concerning the ability to manipulate these cells to enhance a person's capabilities.

What was it like to be a young woman in a laboratory that was cloning some of the first mammals? What were the motivations of the people working there? And why did people want to clone animals in the first place? How did the repeated failure to clone healthy mammals lead to stem cell technology and the banking of umbilical cord stem cells? Clara's chapter 8 is a first-person account of what it was like to be in such a laboratory during this time and how things changed enormously once Dolly the sheep was born.

7

CLONING ANIMALS, CELLS, AND GENES

—

Where Did Cloning Come From, and Where Is It Going To Right Now?

SCOTT GILBERT

The story may seem a bit messy, but that's because life is messy, and science is a slice of life.

—Ian Wilmut (head of the group that created Dolly), 2000

One novel way of reproducing oneself is to be cloned. The result would be a genetically identical offspring. Instead of having only half your nuclear genes, your offspring would have the complete set. (Only the few mitochondrial genes would be different.) However, this is much more in the realm of science fiction than frozen oocytes, surrogate wombs, and other such new procedures. However, cloning has spawned several new technologies, including stem cell transplantation and gene editing, and these promise to have significant effects on our views of personhood and what is natural. The word "clone" comes from the Greek word *klon*, meaning a plant cutting or sprig. A complete plant may be propagated from a single piece of a parent plant; apple trees, for instance, are routinely grown from stem cuttings, and even a small part of the runner of a strawberry plant can eventually produce a whole field of fruit.

The science of cloning animals is based on two principles. The first is that every cell nucleus in the embryo or adult carries the same genes (i.e., the genome established at fertilization). The second is that an egg can be "tricked" into normal cell division and development by procedures other than sperm entry. As we saw in chapter 3, fertilization carries out two processes: the transfer of genetic material and the activation of development. These two events must be done artificially if cloning is to be successful.

EARLY VERTEBRATE CLONING EXPERIMENTS: FROGGY VENTURES

The idea for cloning began with the early experimental evidence that the genome is the same in all of a body's cells.[1] In the 1950s, the laboratory of Robert Briggs and Thomas King (1952) demonstrated that when nuclei from early embryonic frog cells were transplanted into the cytoplasm of an artificially activated egg, the newly implanted nucleus could direct the development of complete tadpoles. However, when nuclei were taken from adult cells (instead of from embryonic cells), they found that this totipotency was limited: they could not get complete tadpoles. John Gurdon's laboratory (1962) was able to get full *tadpoles* from nuclei transplanted from adult cells, but, contrary to numerous science fiction books and magazine articles, a nucleus from an *adult* animal's differentiated cells had never produced another *adult* animal—until 1997, and the arrival of Dolly the sheep.

CLONING MAMMALS

Early in 1997, Ian Wilmut, of the Roslin Institute in Edinburgh, Scotland, shocked much of the world when he announced that a female Dorset sheep named Dolly, born to a surrogate mother in July 1996, had in fact been cloned from an adult cell nucleus taken from an adult female sheep (Wilmut et al. 1997). This was the first time that an adult vertebrate had been successfully cloned using an adult nucleus, an event most biologists had predicted was years away from happening, if indeed it ever proved possible at all.

How did Wilmut and his colleagues "achieve the impossible"? First, they took cells from the mammary gland (udder) of a pregnant six-year-old Dorset ewe and grew these cells in plastic dishes. The growth medium was formulated to synchronize the chromosomes of the cells with those of a selected egg. They then obtained maturing egg cells from females of a different breed of sheep (Scottish blackface) and removed their nuclei. The mammary cells and the enucleated oocytes (those immature eggs whose nuclei were removed) were fused by squeezing them together and sending electrical pulses through them; the electric pulses destabilized the cell membranes and allowed the cells to fuse together. Moreover, these same electric pulses activated the eggs to begin development. The cells divided, and the resulting embryos were eventually transferred into the uteri of pregnant Scottish blackface sheep.

Of the 434 fused oocytes created during this experiment, only one survived to adulthood: Dolly. DNA analysis confirmed that the nuclei of Dolly's cells were indeed derived from the Dorset sheep from which the donor nucleus was taken. None of the genes necessary for development were lost or mutated in any way that would make them nonfunctional. That Dolly was a fully functional reproductive adult was proven when she mated normally with a male Dorset sheep and gave birth to her own offspring.

Dolly, however, was not a totally healthy sheep, and her early death is said to have been caused by her original nucleus being from an adult. She was "born old." Since 1997, laboratories around the world have achieved confirmed clonings of sheep, cows, rabbits, mice, cats, and other mammals. Although it appears that all the organs were properly formed in most of the cloned animals, many of the clones developed debilitating diseases early in their lives.

WHY CLONE MAMMALS?

Scientists working on mammalian cloning were after several applications, both medical and commercial. All things considered, there were perfectly legitimate reasons why the techniques for mammalian cloning were developed first by pharmaceutical companies rather than at universities. Cloning is of interest to some developmental biologists who study the relationships between the nucleus and cytoplasm during fertilization and by some scientists who study aging (and the loss of nuclear potency that appears to accompany it); but cloned mammals are of special interest to those concerned with **protein pharmaceuticals**.

Important protein drugs include insulin (for treating diabetes) and blood-clotting factors (for treating hemophilia). These drugs are difficult to manufacture biochemically and were originally expensive. Some of them can be obtained from animals (insulin, for example, was traditionally obtained from pigs), but because of immunological rejection problems, patients usually tolerate human proteins much better than proteins from other animals. So how do we obtain large amounts of the specific human proteins we need?

One of the most efficient ways of producing protein pharmaceuticals is to insert the human genes that code for the desired protein into the oocyte DNA of sheep, goats, or cows. Such an insertion results in a **transgene**, and the animals containing such gene insertions are called *transgenic animals*. A transgenic female sheep or cow might contain not only the gene for the human protein, but might also be able to express the gene in her mammary tissue and thereby secrete the needed protein in her milk (Melo et al. 2007).

Producing transgenic sheep, cows, or goats is a highly inefficient undertaking. Only about 20 percent of the treated eggs survive the insertion and develop into transgenic adult animals. Of these adult transgenics, only about 5 percent actually express the human gene. And of those transgenic animals expressing the human gene, only half are female, and only a small percentage of these actually secrete a high level of the protein into their milk. Moreover, it often takes years for them to first produce milk, and after several years of milk production, they die, and their offspring are usually not as good at secreting the human protein as the original transgenic animal (Meade 1997).

Cloning would enable pharmaceutical companies to make numerous "copies" of an "elite transgenic animal." Such cloned transgenics should all produce high yields of the human protein in their milk. The medical value of such a technology would be great, and human protein pharmaceuticals could become much cheaper for patients, many of whom depend on them for survival. The economic incentives for cloning were also enormous. Thus, shortly after the announcement of Dolly, the same laboratory announced the birth of a cloned sheep named Polly (Schnieke et al. 1997; Pollack 2009). Polly was cloned from transgenic adult sheep cells that contained the gene for human clotting factor IX, a gene whose function is deficient in hereditary hemophilia.

WHY DO CLONES HAVE HEALTH PROBLEMS?

Many cloned animals suffer from obesity, liver failure, brain malformations, respiratory distress, and dysfunction of the circulatory and immune systems. The questions of why so few cloned animals survive to be born, and why those that do survive tend to have serious health problems may be related (Burgstaller and Brem 2016). First, cloned animals may be prematurely old; that is, the newborn clone's cells may reflect the age of the adult animal from which they were cloned. Second, the DNA of the differentiated adult cells used as nucleus donors is highly modified, and it may be extremely difficult to return the DNA to the unmodified, undifferentiated state found in early embryonic cells. Even though different cell types all contain the same complement of genes, it is clear that not every gene can be active (i.e., produce its protein) in every cell—the result would be molecular chaos.

The major type of DNA modification under scrutiny is *methylation*. In methylation, methyl groups—small biochemical groups made up of one carbon atom and three hydrogen atoms (CH_3)—are added to the DNA molecule, preventing the gene from being activated. Different regions of DNA are methylated in different cell types: In red blood cell precursors, for example, the DNA of the globin genes is unmethylated (active), but the gene for insulin is methylated (inactive). In the pancreatic cells that secrete insulin, the methylation pattern is reversed.

Several laboratories have found that the genes of cloned animals have abnormal methylation patterns (Boiani et al. 2002; Jaenisch and Wilmut 2001). Apparently, while many of genes can be "reset" to their undifferentiated state (a process referred to as **epigenetic reprogramming**), other genes may retain their differentiated methylation pattern. Faulty methylation patterns leading to faulty gene activation (i.e., faulty protein production) would explain why so few cloned embryos survive, and also why surviving clones are plagued with health problems. This phenomenon presents scientists with an interesting question as to whether and how cellular differentiation can be reversed. Indeed, some researchers believe cancer involves abnormal DNA methylation; thus, cloning studies may also provide new insights into how cancer cells arise and grow.

In summary, cloning is not a very effective technology, and it does not often yield healthy offspring.

THE BIOLOGY OF STEM CELLS

Which leads to embryonic stem cells. As we saw in chapter 5, each cell of an early embryo is capable of becoming every type of cell in the body, as well as that of the fetal placenta. These cells—such as the first eight cells of the human embryo—are **totipotent** cells. Once the blastocyst has formed, the external cells become the placenta precursors, whereas the cells of the inner mass have the ability to produce all the cells of the embryo. These inner mass cells are now called **pluripotent** stem cells. It is these pluripotent cells that, when taken from the embryo and grown in the laboratory, are called **embryonic stem cells**.

Stem cells have two critical properties:

1. They have the capacity to divide for indefinite periods of time.
2. They have the ability at each cell division to give rise both to a similar stem cell as well as to a cell that can differentiate into a more specialized cell type.

That is, in addition to generating a more specialized type of cell, stem cells also generate more stem cells (figure 7.1). This is crucial, because it means the population of stem cells is relatively constant, meaning that more specialized cells can continually be made (Martin 1981; Gilbert and Barresi 2016).

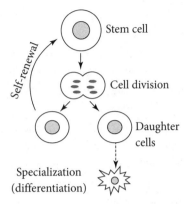

Self-renewal

Stem cell

Cell division

Daughter cells

Specialization
(differentiation)

FIGURE 7.1 **Stem Cells.**

Stem cells have the property of dividing such that one of the daughters remains a stem cell while the other can differentiate into a specialized cell type. In this way, there are always stem cells to continue growth. The daughter cell that remains a stem cell usually stays attached to the "niche" of other cells surrounding it, whereas the daughter cell that differentiates goes outside this niche.

Embryonic stem cells are cells that keep reproducing. Generally, when they divide, one of the cells remains an embryonic stem cell. The other daughter cell can develop into any cell of the embryo, which is determined by what other cells it meets. As mentioned, an embryonic stem cell is a cell that can generate an embryo. It arises from the inner cell mass of an early embryo that does not yet have any recognizable structure. The embryonic stem cells are derived from a ring of fewer than 250 cells.

There are also *adult stem cells*. Our bone marrow, for instance, has blood stem cells, and these are the cells that allow bone marrow transfers from one person to another. These cells are **multipotent**. Whereas totipotent cells can generate the embryo and the fetal portion of the placenta (the mother forms the other half), and pluripotent cells can generate all the cells of the embryo (but not the placenta), multipotent cells can generate many cell types. The blood stem cells of the bone marrow, for instance, can generate red blood cells, white blood cells, and the lymphocytes of the immune system. Interestingly, blood stem cells are also found in the umbilical cord that connects the fetus to the mother. The blood supply and the matrix of the cord provide a rich source of blood stem cells and may also contain another type of multipotent stem cell that can generate muscle, bone, and connective tissue.[2]

The potential importance of stem cells for medicine is enormous, and the potential of the pluripotent embryonic stem cell, a cell that can generate any cell in the adult body, is currently firing the medical imagination. Imagine having a supply of cells that can generate any normal cell type in the body. A few possible scenarios include using human embryonic stem cells to produce new neurons for patients with degenerative brain disorders (such as Alzheimer's disease or Parkinson's disease) or spinal cord injuries, new pancreatic cells for those with diabetes, and new blood cells for people with anemias. In people with deteriorating hearts, it might be possible to replace the damaged tissue with stem cell-derived heart cells, and, in those suffering from immune deficiencies, it might be possible to replenish their failing immune systems.

Human embryonic stem cells can be obtained from two major sources. First, they can be derived from the inner cell masses of human embryos. The source of these cells is typically the embryos left over after IVF procedures, since these procedures generate many more embryos than are actually transplanted. Second, embryonic stem cells can be generated from gamete precursor cells (that would form sperm and egg) derived from fetuses that have miscarried. In both instances, the embryonic stem cells are pluripotent.

Embryonic stem cell therapy has already worked in mice. Mouse embryonic stem cells have been cultured in conditions causing them to form lineage-specific stem cells capable of producing insulin-secreting pancreatic cells, muscle cells, glial cells, and neural cells. For instance, when mouse embryonic stem cells were placed in a dish containing two particular embryonic proteins, the embryonic stem cells divided into *glial* cells, which support the nervous system, maintain neurons, and may play an important role in memory storage. If the same embryonic stem cells were cultured in a medium containing a different mix of embryonic chemicals, they became *neural* cells. Most importantly, these glial and neural cells were functional. When placed into the brains of diseased mice, they were able to restore glial and neural functions. Indeed, neurons derived from mouse embryonic stem cells have been shown to significantly reduce the symptoms of a Parkinson's-like disease in mice (Brüstle et al. 1999; McDonald et al. 1999).

Although human embryonic stem cells differ in some ways from their mouse counterparts in their growth requirements, in most ways they are very much alike. Like mouse embryonic stem cells, human embryonic stem cells can be directed down specific developmental paths. For example, researchers have been able to direct human embryonic stem cells to become blood-forming adult stem cells (called *hematopoietic* stem cells), which could further differentiate into numerous types of blood cells.

INDUCED PLURIPOTENT STEM CELLS

One big difference between laboratory mice and humans is that lab mice are inbred and genetically identical. Humans, obviously, are not. This means that as human embryonic stem cells differentiate, they express significant amounts of certain proteins that can cause immune rejection. How can one get embryonic stem cells that match a particular patient? The breakthrough came in 2006 when Kazutoshi Takahashi and Shinya Yamanaka (2006) showed that there were certain genes that were active in normal mouse embryonic stem cells, and that when these genes were activated in normal adult mouse fibroblasts (the skin cell type that repairs cuts), the fibroblasts would turn into the equivalent of embryonic stem cells. A year later, they and others (Takahashi et al. 2007; Yu et al. 2007) showed that these genes would cause human fibroblast cells to become the equivalent of embryonic stem cells. Such cells are called **induced pluripotent stem cells (iPSCs)**. In mice, these cells can generate all the cell types of the embryo. Moreover, it has been shown that when the natural embryonic stem cells of a normal mouse embryo were removed and replaced with iPSCs, the entire embryo came from the iPSCs. This means that the embryo was derived from a single adult cell that had been made into an iPSC. It also appears that iPSCs remain "young." Their pattern of DNA methylation is very similar to that of a real embryonic stem cell.

The therapeutic potential of iPSCs was demonstrated by the ability of iPSC-derived hematopoietic stem cells to correct a mutant phenotype in mice (Hanna et al. 2007). Here, skin fibroblasts from a mouse having sickle-cell hemoglobin were made into iPSCs by activating the genes that induce pluripotency. These cells were then given DNA containing normal globin genes. These genetically corrected iPSCs were then cultured in media that promoted the production of adult blood stem cells, and these adult stem cells were injected back into the mice with sickle-cell anemia. Within two months after this intervention, the anemia had been cured. This type of curative technology is presently being studied in human patients (figure 7. 2).

Takahashi and Yamanaka received the Nobel Prize for their pioneering work in making pluripotent stem cells. Their technique could be a potential method of both reproduction and regeneration. In the not-too-distant future, spinal cord injuries, liver failure, heart failure, and neurodegenerative diseases such as Alzheimer's disease and Parkinson's disease may be cured through embryonic stem cell therapy.

Hope for such cures was advanced in 2014 when two laboratories (Pagliuca et al. 2014; Rezania et al. 2014) exposed mouse iPSCs to compounds that they would see during development, causing them to differentiate into insulin-secreting pancreatic cells. When scientists placed these insulin-secreting cells into diabetic mice, the mice were cured of their diabetes. Therefore, such cells may generate a new type of therapy for diabetics.

There are several ethical issues involving iPSCs. One such issue involves aging (Gilbert et al. 2005). Can iPSCs delay death? If we can continually replace our aging bodily cells with organs generated from iPSCs, can we delay mortality? Should a person be allowed to live 150 years? What would be the consequences if only a part of the population could afford the therapy that would allow them to live healthy lives past one hundred years? Should we allow this technology to be used in people past 100? What are the financial ramifications of living in a nation where relatively few people work and most are retirees in their tenth decade? When do people retire if they're living healthy elderly lives, and how are their lives supported?

Another set of moral issues involves the new ability of making germ cells—the sperm and egg—from iPSCs (Cohen et al. 2017). If one can make functional sperm and egg from skin cells (something now possible in mice), then a person has the ability to make a baby without another parent. The sperm and egg could conceivably (pardon the pun) come

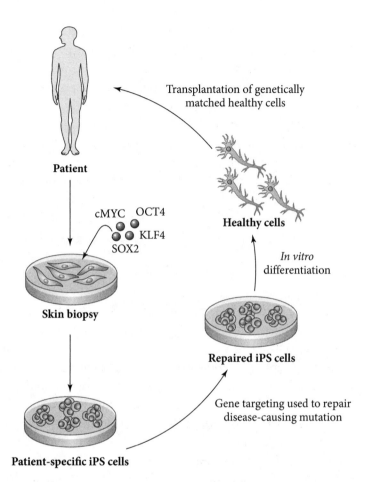

FIGURE 7.2 Induced Pluripotent Stem Cells and their Possible Means of Treating Disease.

Skin samples taken from a patient are treated with viruses producing factors that induce a cell to become pluripotent. These become patient-specific induced pluripotent stem cells (iPSCs) and can be treated so that they differentiate into particular cell types. This type of therapy would replace damaged or degenerating cells. Alternatively, the genes of the iPSCs could be edited and put back into the patient. This procedure could be used to repair genetic lesions.

from the same person. Also, if one could generate eggs from such cells, then there would be no barrier to a woman's having eggs at any time in her life. There would be no need for ovarian hyperstimulation, either. Similarly, sperm could be generated from men whose own germ cells had been eliminated by chemotherapy or radiation.

But we shed skin cells all the time. If these cells could be made into gametes, the theft of genetic material could be enormous. "Imagine," suggests a recent article (Mullin 2017) in the *MIT Technology Review,* "you are Brad Pitt. After you stay one night in the Ritz, someone sneaks in and collects some skin cells from your pillow. But that's not all. Using a novel fertility technology, your movie star cells are transformed into sperm and used to make a baby. And now someone is suing you for millions in child support."

GENOME EDITING AND ENHANCEMENT

Once one has a cell—like an iPSC—in culture, it can be manipulated. For instance, one can edit its genome. Numerous techniques are available for this purpose, but recently an efficient and inexpensive technology called CRISPR (which stands for "clustered regularly interspaced short palindromic repeats" and is pronounced "crisper") has made changing genes relatively easy (Sander and Joung 2014). What makes it efficient is that it has actually been tested for billions of years. CRISPR was not so much invented as discovered. It is a basic part of the bacterial immune system that protects bacteria against viruses.

The major component of CRISPR is a gene that has two critical sites: a guide sequence made of ribonucleic acid (RNA), which can find a similar gene elsewhere in the genome, and an attachment sequence that recognizes an enzyme that cuts DNA (figure 7.3). As a result, whatever gene is recognized by the RNA made from this gene is cut and removed. The secondary component of this system is a revised copy of the gene one wishes to insert. Once the cut is made and a region of a gene is removed, the DNA repair mechanism of the cell will replace it with the new version.

The knowledge of this process has allowed scientists to change the DNA sequences of numerous animals, including mice, by chopping out

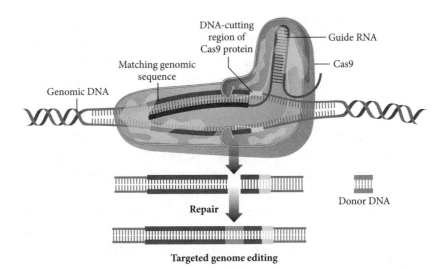

FIGURE 7.3 CRISPR Technology.

The CRISP contains RNA that is complementary to the region of the genome one wishes to edit and binds to the DNA of this region. On another portion of this RNA is a sequence that binds the Cas9 enzyme, which cleaves the DNA in this region and allows a donor piece of DNA to enter. Enzymes that normally repair the DNA will insert this DNA into the region.

specific genes of embryos. Genes have been changed in human iPSCs, and there is one reported case of scientists attempting to alter sequences in a human embryo. However, these last experiments (Liang et al. 2015; conducted on human embryos with chromosome anomalies that would not allow them to survive long) did not work, as the CRISPR produced unexpected mutations in other genes.

But these experiments highlighted important questions: questions about power, responsibility, and who we are (Baltimore et al. 2015; Kaiser and Normile 2015). As more and more genes and their protein products are identified, our knowledge of both genetic disease and the processes of human development becomes greater. With the application of this rapidly expanding knowledge to medical technology comes the prospect of a world free from debilitating genetic diseases. This bright, wished-for world most researchers envision is one in which such human scourges as cancer, cystic fibrosis, and Parkinson's disease can be cured or even eradicated.

But many people see a "dark side" to gene therapy (CGS 2015). The idea of deliberately altering a person's genes seems to some to be beyond the proper scope of medicine. And if the technology should be applied to nonmedical conditions—for example, altering the gene for human growth hormone merely to allow a child to be taller than his or her peers—a whole new spectrum of issues emerges. The term "genetic engineering" can refer to either of two different forms of gene therapy, depending on the cell type being modified. **Somatic cell gene therapy** is essentially a medical treatment intended to target abnormally functioning genes in specific somatic (bodily) cells of a single patient. **Germline gene therapy** is intended to modify all of a person's cells, including the sperm and eggs. Not only is the person's genome altered by germline gene therapy, but the descendants of the person can inherit the altered genes. It is germline gene therapy that generates the most wide-ranging ethical questions.

SOMATIC CELL GENE THERAPY

In the most common form of somatic cell gene therapy, a normally functioning gene is inserted into a patient's genome, usually through a virus, replacing a poorly functioning gene. The hope is that the inserted gene will be accepted by the cell and translated into a normal gene product (a protein), thus relieving the patient's suffering. Because the new genes are not inserted into the patient's germ cells, no genome changes are passed on to offspring. Somatic cell gene therapy is therefore seen as a medical intervention, like heart surgery or radiation therapy, designed to cure or alleviate disease in a single individual.

However, this type of gene therapy is still experimental, and it has some technical problems. The major problem is the ability of the virus carrying the new gene to integrate into the genome effectively. Another is the regulation of the introduced gene. If its product is made in too great a quantity, the consequences can be dangerous.

The combination of CRISPR technology and stem cells can come together to get around these problems. If the genes of a person's stem cells can be isolated and modified by CRISPR, then just the targeted mutation is corrected, and the regulation of the gene continues to function normally. The stem cells can be put back into the patient. This is still an experimental technology, and the efficiency of gene editing and gene delivery and the specificity of the CRISPR proteins still need to be improved (Cox et al. 2015).

Most scientists and medical professionals agree that treating a disease by inserting a corrected gene into a patient is not ethically different from using medicines to treat disease. If sickle-cell anemia or Huntington's disease could be cured by administering effective drugs, doctors would certainly do so; likewise, if they could be cured simply by adding the normal form of the affected gene to a patient's blood cells or neurons, few would quibble. Most of the concerns voiced about somatic gene therapy are the same as those heard about any cutting-edge medical advance: the safety of the procedures and the equity of their availability. In addition, however, gene therapy raises the issue of medical enhancement (as opposed to medical treatment).

The boundary between treatment (for a disease) and enhancement (for cosmetic or athletic purposes) is indistinct. For instance, is short stature a disease? Statistics would argue that taller people have a better chance of success than short people. Is infertility a disease? It prevents an individual from reproducing but does not usually cause harm in and of itself. Many conditions not usually considered to be "diseases" are treated by the medical establishment, and some, such as baldness, are even covered by health insurance plans.

Enhancement through gene therapy could conceivably become a huge business. In the United States, the controversy over the use of dangerous, illegal steroid drugs to enhance athletic performance shows how great the desire can be for enhancement. Some athletes are willing to risk severe potential side effects and legal prosecution in order to gain the added strength and power these drugs offer. Gene therapy offers the possibility of enhancing strength and athletic performance without the risk of drugs, or even the effort of exercise.

For instance, all vertebrates, including humans, possess a gene that encodes the protein myostatin (McPherron et al. 1997). This protein is a growth regulator that signals muscle cells when they have reached the proper size. Studies in mice show that if the gene for myostatin is absent, the mice grow into muscular rodents who are stronger and faster than their nonmutant littermates. In these studies, individual muscles from myostatin-mutant mice weighed two to three times more than muscles taken from normal mice; the increased muscle mass resulted from both an increased number of muscle fibers and an increased size of individual fibers. One of the discoverers of this gene commented that the myostatin-deficient rodents "look like Schwarzenegger mice." The news media printed articles about these mice, immediately linking the story to what athletes might do if such therapy were available in humans. Indeed, a human child has been found who is deficient in this gene. He has exceptional musculature and is much stronger than other boys his age (Schuelke et al 2004). Although he is healthy, physicians are watching him closely because the same gene is active in heart muscle, where such enlargement can be dangerous. Can human athletes be "constructed" to have such a genome? The answer is not unimportant. A laboratory in China (Zou et al. 2015) has recently announced the creation of large-muscled myostatin-deficient dogs, produced by CRISPR technology.

Whatever treatments or enhancements somatic cell gene therapy might provide, however, would be limited to a single treated individual. *Germline* gene therapy holds the prospect of genetic alterations that could be passed along to future generations.

GERMLINE GENE THERAPY

In germline gene therapy, sometimes called **inheritable genetic modification (IGM)**, the goal is to alter the genome at the germ-cell level, so that the corrected or enhanced gene is transmitted to the person's offspring. Modification of sperm and egg genes has become a routine procedure using laboratory mice and has contributed much of our knowledge about the actions and interactions of many vertebrate genes during development.

Whether this technology should be applied to humans has caused considerable debate (see Harris and Darnovsky 2016). There are strong advocates of furthering the fledgling technology (Stock 1999; Harris 2010). James Watson (2000, 29), one of the discoverers of the structure of DNA, has opined, "If we can make better human beings by knowing how to add genes, why shouldn't we do it?" But many others believe that germline engineering research should be strictly regulated or even banned. Given that germline gene therapy might be able to eradicate inherited genetic diseases and enable us to expand our genetic repertoire, why should anyone be against it?

In fact, many people question whether the therapy is even needed for medical purposes. The American Association for the Advancement of Science (2003) has indeed identified a few specific instances in which the germline modification approach could be used to prevent parents from transmitting defective genes to their offspring. But several alternative procedures—including prenatal genetic diagnosis, gamete donation, embryo selection, and adoption—are currently available and would not evoke the issues involved with germline manipulation, so there might not be a great medical need to do it. Other arguments against IGM range from the pragmatic to the ethical, moral, and emotional.

One safety issue that applies to both somatic and germline therapy concerns the way in which the corrected DNA is inserted into the target cells. These methods often use viruses to transfer the DNA into the cells, and such "viral vectors" can trigger massive, systemic immune responses. Indeed, this happened in one of the first tests of somatic gene therapy in humans. Here, a man who was treated for a defective liver enzyme died from the results of a massive immune response directed against the viral proteins (Wilson 2009). Another argument used against IGM is that when altered genes are edited and inserted into the genome, they may disrupt presently functional genes and cause mutations. This has certainly been encountered in laboratory mice. In one case, the disruption of a single gene resulted in mice that were born without eyes, semicircular ear canals, or a sense of smell (Griffith et al. 1999). In another case, a transgenic strain of mice developed normally but had a high rate of liver cancer owing to the malfunctioning of other regulatory systems (Leder et al. 1986). In the one reported case of using CRISPR on human embryos, numerous mutations, including changes in genes needed for vision and cell adhesion, resulted (Liang et al. 2015).

Proponents of IGM point to advances in targeted gene insertion as an indication that stumbling blocks may yet be overcome. For instance, advances have been made in CRISPR technology since the original 2015 paper on human embryos. However, in the case of inheritable manipulations to the germline, some effects may take several generations to manifest themselves—and any mistakes made will be permanent. Inheritable gene modification is not a drug that can be discontinued if the side effects are disastrous (Newman 2003).

A second argument often heard when a new technology is introduced is that "we're playing God." This argument was used against Benjamin Franklin's lightning rod, and it was used against smallpox vaccination (White 1896). This is akin to the emotional argument that says we should not interfere with nature. But medical interventions are never undertaken with the view that nature is totally benign, and they are by their very existence interfering with nature. One can—and some do—accuse heart surgeons who perform bypass surgery or neurosurgeons who remove brain tumors of playing God. Most of us, however, are glad they do. Indeed, in some religious traditions, "playing God" by healing is considered a most worthy endeavor.

A third argument against this technology (and others) is that "we do not know what such genetic technology will be used for." If lethal genetic diseases such as Lesch–Nyhan syndrome or Huntington's disease can be screened for by preimplantation genetics, germline genetic engineering becomes a very high-tech solution for a problem that has a relatively low-tech cure. So what else might the technology be used for? One possibility is that it could be used for phenotype enhancement.

Modern plastic surgery has allowed thousands of people to live better lives. The ability to restore facial or limb structure and function to those who have lost it as a result of trauma or genetic malformations is one of the highest achievements of modern surgical technology. In addition to such cases, however, millions of people make use of these same surgical techniques to make themselves more attractive. Similarly, medical advances used to fight prostate tumors can also be used to prevent male baldness. It is reasonable to expect, then, that genetic technologies designed to fight diseases will also be used for purposes of enhancement. We know there are genes that affect height and muscle mass, so we could conceivably make our offspring taller and stronger. If genes involving intelligence were found, those who could afford this procedure might enhance themselves in the hopes of producing highly intelligent offspring.

Whereas somatic cell gene therapy is like any other medical proce-dure in providing value for the patient, IGM differs in providing value for the patient's offspring. Jeremy Rifkin (1998) voices his concern that "those families who can afford to program 'superior' genetic traits into their fetuses at conception could assure their offspring an even greater biological advantage, and thus a social and economic advantage as well." Lee Silver (1998) envisions a world where, owing to economic inequality, the genetic "haves" and "have-nots" are far apart in their abilities: Genetic engineering would convert economic differences into inherited biological differences (Chapman and Frankel 2003). At the moment, there are no restrictions on what such therapy could be used for. Unless worldwide legislation can define what is and is not allowable, this remains an impor-tant critique of the technology.

A related argument is "do we really know which traits to enhance or get rid of?" Genetic engineering assumes that we know which traits are good and which are bad. However, what is good in one environment might be harmful in another. The genetic mutation that causes sickle-cell anemia is harmful when homozygous (that is, when it is inherited from both parents)—but when only one copy of the gene has the mutation, the overall effect may be advantageous in certain environments since it may offer some protection against the parasitic disease malaria (Weiss 1998). A mutation in a gene for a certain molecule on lymphocytes may normally be a bad thing, but this same mutation may offer protection against the human immunodeficiency virus (HIV), the virus that causes acquired immune deficiency syndrome (AIDS). Similarly, the same muta-tion that predisposes people in Western populations to allergic reactions and asthma is advantageous in areas of the world where certain parasites are a major health problem.

If we were to know which genes or groups of genes produce aggressive or docile phenotypes, should we change them? Are certain gene vari-ants or combinations of gene variants deleterious in some situations but predisposing toward acts of genius in others? We just don't know. There exists a fear that the traits people choose for their children will not be healthy in the long run. Some question the consequences if a trait chosen in one generation falls out of fashion in the next or becomes particularly ill-suited to a change in the environment.

Similarly, one might ask, "Do we even know the functions of the genes that might be changed?" It is one thing to look at genes for proteins that are the end-products of development—hemoglobin or insulin, for example. These genes probably have a single function. But genes that act during development often have many functions, a condition called **pleiotropy**. Expression of the *BMP4* gene, for example, can induce bone growth in some tissues but induces apoptosis (the death of cells) in a different set of developing tissues. In ectodermal tissues, the same *BMP4* gene product can induce cells to differentiate into skin epidermis instead of nerves. We are constantly discovering that genes are not "for" a particular function; rather they are "used in" a particular function. If we alter a gene in order to affect one function, we may well find that it also disrupts another function.

Also, one might criticize IGM technology by asking, "Do parents have the right to make decisions about their children's genotypes?" In the normal scheme of things, there is a great deal of chance involved in which traits a child will inherit from his or her parents. But what if the child's genes were "designed" and paid for by the parents? The parents would directly control the qualities of their offspring. If the inheritance of certain traits were a certainty, the individuality of the child could be affected. If parents were to select genes for height and body musculature, they might then pressure their child to succeed at sports, regardless of whether the child *wants* to play the game. The entire notion of individual personhood is called into question. These questions are similar to those raised by the issue of reproductive cloning.

A range of critics believe that germline genetic engineering could convert a child into a commercial product with expected parameters of normalcy and function. Such critics maintain that we might end up seeing "fads" in children—one generation preferring a certain hair color, height, or organ endowment in its children—and the standards for what is genetically desirable would likely be those of the society's economically and politically dominant groups. People who fell short of some technically achievable ideal would be seen as "damaged goods," increasing prejudice and discrimination.

Disability rights advocates are critical of germline engineering technology because they fear that a social objective of establishing the "perfect" human might lessen society's value of care and respect for all human beings. In addition, the loss of care and respect for the less fortunate would leave people with disabilities as pitied mistakes, born with genetic diseases that could have been corrected.

It should be noted that not all genetic diseases are foreseeable. The notion of a world (or a country) free from genetic disease is not going to happen. For instance, the most common form of dwarfism—achondroplasia—is caused by a dominant genetic mutation that causes cartilage in the arms and legs to stop growing too soon. About seven out of every eight cases of achondroplasia are not inherited but a result of new mutations, mostly carried on the sperm. These are random events that cannot be predicted. (And many people with achondroplasia do not find their condition to be a "disease.")

Most scientists believe that *every* human being is a heterozygous ("recessive") carrier for several harmful genes. In other words, every one of us carries one "bad" copy of at least a few genes. This is why marrying close relatives is a dangerous enterprise. In most instances, an unrelated couple will not both carry bad copies of the *same* gene. However, if close relatives marry, the odds rise that they will carry the same mutant recessive genes and that their offspring will be adversely affected.

One last criticism of genetic germline modification, as we have already seen, is that such genetic engineering may lead to eugenics, the late-nineteenth century program that advocates breeding better humans (like breeding better crops and livestock) and was considered a major scientific area until the end of World War II. Eugenics attempted to make the human race more uniform and healthy (Eugenics Archive 2016). While the historical movement was based on unsound biological principles, eugenic goals might now be able to be achieved through biotechnology. But such engineering of the genome might have consequences in reducing biological diversity. Human diversity has been important in resisting disease and certainly makes the world an interesting place. Moreover, as geneticist Theodosius Dobzhansky pointed out, the world's problems are not due to a preponderance of genetically enfeebled people, but to people with excellent genetic endowments using their remarkable brains for antisocial purposes.

Our social problems cannot be underestimated, and there are many who wish that such problems could be cured by genetic means. This is a dangerous line of thought. For instance, communities of color have historically suffered from the racist social applications of eugenic theories, and therefore many modern black leaders are concerned that germline engineering offers another opportunity for racism to manifest, veiled as science. Although most scientists involved in germline engineering have no explicitly racist agenda, civil rights advocates have found it disconcerting that David Duke, former national director of the Ku Klux Klan, heartily supports inheritable genetic modification development. The case for genetic engineering hasn't been helped by those scientists who attempt to promote it by claiming it will "cure" homosexuality, criminality, and homelessness. In the past, the concern was that eugenics would arise from government policy (as it did in Nazi Germany, with historically horrifying results). However, such policies could also come from social pressure and economics.

8

GLORY DAYS

———

My Personal Account of Cloning

CLARA PINTO-CORREIA

Without deviation from the norm, progress is not possible.

—Frank Zappa

Sometimes, when I look back at those long-gone days of my youth, I still have trouble believing that I was a postdoctoral fellow[1] doing mammalian cloning exactly when Dolly was born—in the winter of 1996, in what now seems to be a galaxy far, far away. I can still see our faces as we rushed to the lab early in the morning, hardly believing what they kept repeating on the radio with such enthusiasm. I still remember looking around and knowing right away that we were all in this together. We didn't even know what to say.

Everywhere in the world, at that precise moment, all the media, in all languages, were telling people history had just been made because a cloned sheep had been born.

That sheep, they insisted, was the first cloned mammal ever born on earth. They were all up in arms. And they all seemed to us to be screaming rather than talking.

We tried to talk to each other by starting some sentences, but we were all so stunned that we never got to finish them, at least in any coherent form.

Yes, there was a strange confusion; and sure, this confusion was universal. We were quite an international crew, so we knew they were repeating the exact same blunder, at least in Portugal, France, Argentina, India, Norway—even in the heart of Kansas, where our boss came from. The whole thing was so truly weird. Those people were all saying that Dolly was the first mammal on earth ever cloned by human ingenuity. But no, we knew she wasn't. She was just the first mammal cloned from an *adult* cell. We had already spent years cloning other farm animals by then. OK, we used a slightly different technique, but still, they *were* mammals. A huge gray cloned female rabbit called Douce lived in my house as a pet, since she was no longer needed for any experiment. And I made no secret of it. Actually, I was so proud of my achievement and she turned out to be so sweet that I talked about her to all my friends and family, and paraded her immediately when I had visitors. But now those multilingual announcers were protesting that this scary sheep had been concocted in total secrecy.

Hold it right there. Scary? Why would a sheep be scary?

Hold it again. Secret? Why would anyone clone a mammal in secrecy?

Yet everyone dispensing news to the world that morning seemed to agree Dolly was scary, and one of the reasons why she was scary was because she had been secretly prepared, God knows what for.

No one had ever told us we were scary or secret or both. We kept staring at each other, not knowing what to do next. We were suddenly discovering that the idea of cloning mammals was a very public, if not a very scientific, media concern.

And we were going to have to deal with all this, too.

All of a sudden, we were surrounded by media stories about clones having belonged solely to science fiction, never being there for any good reason to start with, and not being all that welcomed as real entities invading the real world. But why were they lying so shamelessly? Why didn't they ask someone who knew better? Jesus, some student would suffice, at least some former student of mine. My old embryology textbook from my days as an embryology student had a long section dedicated to Gurdon's cloning experiments with frogs carried out in the 1960s, successful to the point of having cloned embryos that developed into live tadpoles. After my graduation, I had spent several years teaching just that in medical school. I used the first edition of Scott's *Developmental Biology* (we referred to it as "the Bible") to prepare my classes, and it said just what we were saying: Ever since Gurdon, there had been plenty of cloned vertebrates in the world.

So what was this all about, all this talk of danger coming from there never having been a real clone before Dolly?

Absolutely taken aback by the sharp contrast between the sweet routine of our lab life and the incredible fears of the public, I dropped my initial arrogance and just tried to understand the wave of panic. Whatever was going on was strong enough for people to feel that they were being robbed of their most sacred rights in what came to the understanding of life. Cows, goats, rabbits, a pig, and numerous other copies of my beloved Douce had been patiently cloned in one lab or another all over the world since the 1980s. My colleagues had been publishing their results in open-access public journals the way any serious scientist does. The techniques for transplanting the nuclei from one cell to another and for activating the eggs had been openly debated at several meetings all over the world. But none of this seemed to matter now. We had obviously done something wrong when communicating our results, and somebody else had obviously also done something wrong when publicly announcing the birth of Dolly. It had to be, or else our entire known world would not act that profoundly shaken by Dolly's birth. For from what we could hear and read, people were responding to a mere cloned sheep as though something absolutely new, and certainly more threatening than whatever biotechnology had yet been invented, had suddenly entered our lives and was now planted there as the most serious of all living menaces.

This spectrum of public fear and anxiety deeply perturbed what used to be my nice and easy communication routines.

I had been both a biologist and a journalist since 1980. Therefore, when 1996 came around, popular science had long been my middle name. Still, Dolly's birth forced me to learn how to explain cell biology to the general public at a level and intensity that had never been needed before. For the first time in my life, when I gave talks open to any interested person, people insulted me or begged me to stop what I was doing before it was too late. I saw students, already two or three months into their cloning-related research projects for advanced degrees, decide they were going to dedicate their energies to some other topic, always with the claim that this one was going nowhere. A really bright young woman from Mauritius, already having written two papers as first author, a gorgeous kid with such good hands and amazing vision she could easily have become America's cloning queen and who had a cloning company ready to give her a green card even before she finished her PhD, sought refuge in my office. She asked me to lock the door and cried on my shoulder while she poured out her despair. She was quitting cloning because her family back home couldn't take it. Their wonderful daughter couldn't possibly be working for the ugly head of evil.

We certainly played a guilty part in this mess. Before Dolly was born, we had never felt the need to say a word about our work to people other than ourselves. Cloning became badly misrepresented, and some of our best students were being hurt by our carelessness. Shame on us.

THE FIRST INCENTIVES DRIVING MAMMALIAN CLONING

I was doing my postdoc in mammalian cloning at the University of Massachusetts, Amherst, with professor James M. Robl, one of the greatest pioneers in the field. I couldn't have been luckier. Scientifically speaking, the time I spent in Jim's lab was the most productive and gratifying of my life. It's just that, when I first got there, met my colleagues, and we all started greatly enjoying ourselves with the fun side of science, none of us had any idea of what we were getting ourselves into. We could understand our own experiments, but the big picture was rather unclear. As we went to the lab bench to do our work, we also kept going back to Jim's office for ever more questions.

Why would anyone want to clone a rabbit in the first place? As Jim always explained with saintly patience, rabbits were models: They were only meant to allow us to understand enough about mammalian embryo transfer so that we could move on. Yes, boss, but why do we want to clone mammals? Well, apparently, back then, there were two main reasons.

Finding Keys to Cell Behavior

The first reason had to do with basic science, and it dealt with a fundamental question of life: How do the inherited genes turn one cell into a brain cell and another cell with the same genes into a liver cell? And also, quite simply: How powerful could the egg's cytoplasm be to make our transferred nucleus act like a fertilizing sperm, and at which stage of the cell cycle could the egg's cytoplasm be so powerful? Cloning could help answer both these questions, and maybe other questions that we had not yet anticipated. The answers to those questions might eventually become relevant to curing cancer,[2] but for us they became mainly a possible way of explaining that ancient and exciting mystery—how does an egg develop into an embryo?—an answer sought for its own intrinsic value, like the Holy Grail.

Making Expensive Drugs Low-Cost

The second reason for wanting to clone mammals, Jim said, had to do with the future commercial and medical applications of cloning. We were kids and didn't get it yet, so he got professorial on us. There are protein drugs, he explained, such as insulin and clotting factors, which are badly needed by humans—and yet are very expensive and difficult to make or obtain from nature. It would be easier to access them if we had, say, a sheep with a human insulin gene inserted into its genome in the right place, so that it would secrete the insulin into its milk (discussed further in chapter 9). This had been done before, but the result had been inefficient. Some sheep produced lots of insulin; others produced only small amounts. It would be wonderful if the high-producers could be cloned so that we could always count on a large and precise insulin harvest. Got it, boss.

Agriculture and Environment

None of this sounds all that grand, but it was exciting enough for us. Quite obviously, we didn't believe or think seriously of the horror-movie nonsense, and we had not joined the ride at UMass so that our nutty professor could generate a thousand copies of supermodels or United States Marines, or make genetically identical clones of great thinkers locked inside test tubes, awaiting their time to save the world from chaos. We were scientists, not science fiction authors. Our partners were developmental biologists and pharmaceutical company executives. Eventually, when our techniques became more reliable, mammalian cloning would also be of interest to the farming industry, especially where cows were concerned. One of the graduate students on the team wanted to learn mammalian cloning to preserve endangered species. He actually had a project with South Africa concerning the cheetah, which had suffered such intensive inbreeding due to habitat loss and fragmentation that it was becoming infertile. Another young scientist wanted to clone a bull whose meat the Japanese gourmets considered beyond exquisite. And that was just about it. Or so we thought. It was the early 1990s, when we honestly believed that no one in the real world would ever care about our work. We cared enough, we discovered enough new awesome details every week, we worked hard enough night and day, that had we been gamers back then we would have always been on the brink of an epic win, and that was reward-ing enough for us. Yes, we could easily imagine how our efforts would give rise to a grandiose dream. But, for a bunch of young students from a small-time lab like ourselves, this dream would certainly blossom only a long time down the road, probably after our own lifetimes. Except for Jim, we all tended to assume this was the case.

MAMMALIAN CLONING'S MAIN SHORTCOMINGS

This was cloning's age of innocence. We had plenty of complex prob-lems to solve, and so had everyone else in the field. A good number of cloned cows, sheep, and rabbits had already been born since the mid-1980s through the efforts of several different teams. But these were from embryonic cells, not from adult cells. In other words, we couldn't get a high-insulin-producing sheep and clone her yet. We always stumbled on the same opaque limitations. First and foremost, the vast majority of our embryonic clones died shortly after we put them in the incubator, or else after we transferred them to the womb of a "pseudo-pregnant" mother; that is, a female carefully treated with hormones to be exactly at the stage of the cycle when the embryos nest in the endometrium. The magnitude of this massive death was truly stunning: out of three hundred clones, only one succeeded; out of five hundred clones, only two succeeded; out of two hundred clones, only four succeeded; out of five hundred fifty clones, none succeeded, and so on. There never seemed to be any particular reason for this. It just kept happening all the time, to all of us, no matter how many cloned embryos we prepared exactly at the same time, in the same way, and under the same conditions.

Another weird, recurrent symptom in our few successful clones—actually one my huge Douce could have suffered from and which caused her early death—was what we then simply called the "large-calf syndrome." A good number of clones were simply born too big for their own good. This caused them to develop cardiac complications, together with bad bones and bad lungs. These specimens always died young. And although we could easily speculate about what caused this condition, we didn't really have a clue as to why.[3]

MAMMALIAN CLONING AFTER DOLLY

In 1996, the birth of one single sheep put an abrupt end to our quiet lives. In the wake of the euphoria over Dolly, the cloning company Jim had dreamed of for so many years was finally created at UMass. That company produced six transgenic cloned calves one year after Dolly's birth. At that point, the whole veterinary and animal sciences department became an endless media circus. We just didn't know what to do with all the phone calls, both local and worldwide, asking about biological mysteries at a time when the Internet was just in its beginnings and only a few nerds knew how to use something new called email.

Amazingly enough, we all noticed right away that the big scientific breakthrough achieved by the birth of Dolly was completely ignored by the news, although we kept trying to convey it to anyone willing to listen. Before 1997, we used embryonic cells retrieved from blastocysts as the sources of nuclei for our clones. Our rationale was that only those cells were "plastic" enough to be reprocessed by the egg's cytoplasm and start development from the beginning. This meant, however, that we could never know exactly what our clones would be like—we just knew the populations they came from and had to take a guess and live with whatever grew. Dolly changed everything and gave us a great lesson in cell biology. For the first time, an animal had been cloned from an adult, in

Dolly's case using cells from the mammary gland (and yes, her name was coined after the well-endowed country music singer Dolly Parton). This meant that even perfectly well-determined, specialized adult cells could be reprocessed by the egg's cytoplasm to go back to being like a zygote and start the embryonic development of a new organism all over again. And this was quite unlike everything we thought we knew about cell behavior. Time to change the textbooks once more.[4] The cell biology of embryos had completely surprised us again.

For better and worse, Dolly is also a great example of the interactions between *basic science* and *applied science*. This is yet another debate that has remained urgent for decades without any sort of solution in sight. Basic science is research based on curiosity, or even aesthetics. Its goal is to reach the next pinnacles of new knowledge. In basic science, this is done in the form of a publication—and this means, literally, that we scientists make our new knowledge *public*. In applied science, on the other hand, the goal is control and profit. Say you discover a new system to cure a hitherto incurable disease. You gain your profit through the registration of a patent, meaning that the knowledge is owned privately and controlled by its owner. As is easy to imagine, sometimes these goals cooperate (finding new knowledge and helping people at the same time), and at other times they openly conflict (the ownership of a technique or drug might make some people wealthy by limiting its use to those who can afford a large fee).

This division between basic and applied was not usually sharply black and white. The polio vaccine was definitely the outcome of applied science, but its aim was not corporate profit. Jonas Salk and his colleagues simply used the basic sciences of virology and immunology to create a vaccine that saved thousands of lives and changed the way we view childhood disease. However, when asked who owned the patent, Salk famously replied, "The people, I would say. There is no patent. Could you patent the sun?" And, even before Salk, Johns Hopkins biologist George Gey, the man who created the first truly "immortal" cultured cell line in which to develop drugs, got his miracle tools to colleagues as quickly and as healthily as he possibly could without thinking of making money from his success. Indeed, it was in those cells that Salk produced his polio vaccine.[5] Similarly, Wilhelm Röntgen refused to patent his X-ray device.

PUBLICATIONS

If you are a basic science researcher in the area of fertilization and early development, let's assume that all you have is a modest grant to study some mystery within your field of expertise. Therefore, you are not seeking results directly meant to achieve profitable practical applications: You are patiently collecting the bits and pieces of data that you hope will one day allow something very important and very new emerge, change the textbooks, and, maybe, yes, change the world. Basic science might seem like a waste of time and money, but it might also lead to knowledge that can save your life. For instance, in the mid-1980s, it was by studying the basic behavior of fluorocarbon chemicals that a previously unknown chemist, Mario Molina, discovered the hole in the ozone layer that might have been increasing cancer rates enormously.[6]

The science of cloning was initiated by people studying frog and salamander embryos, asking the question, "Does the nucleus of a single adult cell retain all the genes needed to form every cell in the body?" The science behind much cancer research concerned the identification of the genes controlling cell division, based on research carried out in yeast and clams.

Basic science results have another prerogative that makes them extremely important for the progress of worldwide research: They are submitted to peer review, are reworked and rewritten until colleagues and publishers are satisfied, and are published in credible journals—after which, everyone can read them. Moreover, it is de rigueur that if new reagents were used that are not yet available on the market (say, a brand-new antibody made all the difference), the research team authoring the article has to make them available to other teams considering giving them a try.

PATENTS

Patents are a completely different story. Within the confines of academic life, where academic freedom should be the most sacred of all golden rules, a good number of researchers in fertilization and early development have increasingly been patenting their results simply *to make money*. If anyone else wants to use what has been patented, generally for pharmaceutical purposes, it's going to cost them, and the scientists' incomes will seem less meager. Moreover, a corporation's responsibility is not to the public, but to its investors. If a corporation hits the jackpot, it is certainly not going to publish its results in a scientific journal for the entire world to read. The scientists register the patent, the whole technique becomes their intellectual property, and anyone wanting to gather information on even the slightest detail will have to pay for it.

Of course, this is not how science is supposed to function. But, it is more and more how it often does. Reproductive technology has a long history of private knowledge. The best example is the obstetric forceps used to deliver babies who get stuck during delivery. This instrument was invented in the 1600s by the Chamberlens, a family of physicians in England, and it remained their family secret for over 150 years. During all that time, their successes in difficult deliveries enabled them to become extremely wealthy, and allowed them to become the court obstetricians to the queens and princesses of Great Britain. Also during all that time, no one else could safely perform such difficult deliveries, and as a result, women and babies died (Das 1929).

PUBLIC FEARS

For those of us who could immediately grasp the difference between this one cloned sheep and all the mammalian clones we had been preparing earlier on, dreams were coming true, and the future seemed promising. We just needed to make sure that the result everybody was talking about was not simply a fluke that happened to create one single sheep. If Dolly was for real, and if, therefore, the nuclei of adult cells could be used to make new adults, then we could always know from the start what our adults would look like and be like. This was really cloning as it had been dreamed of. We were young and happy, looking forward to a great future.

However, it didn't take us long to notice all the background noise spelling out terrible stories of universal damnation.

It seemed to us a bit strange at first, but yes, our trouble in making ourselves understood was not just the persistent lack of interest in what had changed in the way we understood cell biology. There was more to it. As time passed, people remained anxious—real anxious—about that damned sheep. If our plan B was just to wait for the dust to settle, it was starting to look like it was going to be quite a long wait.

Once more, what had gone so wrong? When? And where? Once more, was anyone prepared to be bombarded with overnight news of a clone from "nowhere"?

Test-tube babies were child's play compared to clones.

The sudden parading of Dolly before a totally unprepared media stands out to this day as a dramatic warning against engaging in self-promotion and science promotion without having first laid the groundwork. Just about anyone who had been serious in their attempts to explain their science has horror stories of trying to explain cloning to a public after the field had been demonized. Eventually, after the hostility and anxiety of trying to explain cloning in popular media, many scientists just stopped trying. It seems that only those with commercial interests continued to talk to the media. Getting wealthy and making medicines were motivations the public accepted. In the end, nobody in the field bothered to explain all that much to the media, nobody in the media was ready to convey the message to the public, the public had been scared for decades, and the result of the ensuing confusion was profoundly damaging.

Scientists should have known better. In 1978, they had watched in total astonishment the readiness with which the public and the media alike reveled in the wild tale engendered by freelancer David Rorvik. In his book, *In His Image: The Cloning of a Man* (1978), a millionaire named Max, intent on becoming his own heir, calls upon a weird scientist named Darwin, who is willing to accept Max's fortune in exchange for producing Max's clone in a secret lab outside the United States. A sixteen-year-old virgin accepted the responsibility of carrying the embryo. This instant bestseller was taken as a true story until a congressional hearing full of testimonies by respected experts in the field managed to dismiss the whole thing as a hoax. But had anyone in science clarified what cloning could ever truly do? Ordinary people know only what they are exposed to, and they knew that cloning had never been a good thing. The best-sellers *The Boys from Brazil*, by Ira Levin (1976), about cloning Hitler; and *Joshua, Son of None*, by Nancy Freedman (1973), about cloning of

John F. Kennedy; as well as *In His Image*, gave us the picture of the mad scientist at the heart of misguided experiments. Michael Crichton's *Jurassic Park* (1990) gave us a story of the idealist eventually being taken over by commercial interests—and then, sure enough, everything gets out of control. People were profoundly perturbed as soon as they heard of Dolly. Interestingly, you could tour the entire world, and the fears would always be the same.

A very common belief was that cloned animals (and real people, why not?) grew inside test tubes and only came out when needed, already grown up and ready for action. Another common fantasy was that mad dictators could now clone their own armies, with the help of not more than a couple of experts and some shrewd viziers. Also, any lascivious sultan with money to spend could clone himself an entire harem of Alessandra Ambrosios—or some other gorgeous fitness queen from Victoria's Secret's Angels, since different people have different fantasies. These reveries are absolutely impossible to achieve in real life, one should note. The first common belief is totally absurd: no clone comes out, all grown up, from any kind of a test tube when we want him to. The dictator wants an army? The sultan wants a harem? Sure, they could get life cells from some Olympic wrestling champion, or from Alessandra herself, and transplant them into hundreds of eggs, all transplants being successful and all clones entering development. Please notice that, in this description, we're assuming that somehow we have finally managed to overcome the massive death rate of cloned embryos (we're not there yet—at all) and, therefore, all the cells transferred to pseudo-pregnant surrogate mothers would have formed nice blastocysts that nested in the uterus and started to grow, leading to a healthy live birth nine months later (as we've already seen, this is also not the rule in any sort of pregnancy).

At this point, both the dictator and the sultan have already waited nine months. But now they are going to have to wait much longer: Those newborn boys need to be breastfed, rocked to sleep in cradles, learn to speak, learn to walk, attend school, go through all the activities of life that make them boys, and then young men—and only then can they face the drill sergeant at boot camp and start being organizing into their deadly squads. We can assume that meanwhile, the dictator has lost the war. Or simply died in his sleep without having invaded all the neighboring countries he dreamed of incorporating into a vast empire. The same with the lascivious sultan. Will he still be sexually ready for all his dazzling Alessandras when they finally grow old enough for the part? Will he even still be alive?

This first quick look at the myths of human cloning brings up yet another extremely important aspect. Exhausted mothers working full-time at high-profile jobs often tell me, not exactly joking, "Won't you please make a clone of me to take care of the kids? That would be so nice." And the reason this question makes no sense is still the same. Clones are not grown inside test tubes and then frozen there, waiting for us to animate them when we need them. A cloned embryo gestates and is born through a pregnancy like any other. Therefore, as much as I can understand how desperately any hardworking mom would enjoy having her own carbon copy so that one can take of the kids while the other one goes to work, and as much as I would hate to let moms down—the truth remains that cloning my friends won't help them.

Imagine I were stupid enough, or just momentarily blinded enough by the heartfelt plea of a woman whose exhaustion I know all too well, to go ahead and grab some adult cell from her body, enucleate some woman's egg, produce the clone my friend asked me for, and softly place it in an incubator. Even assuming that the conditions are optimal and all goes well with the first steps of embryonic growth in the incubator, with the delicate moment of embryo transfer to the uterus, and then with the entire length of pregnancy, the clone born is still just a baby. This baby certainly has the tired mother's genotype, but she will require diapers and breastfeeding, and my friend might be in her early forties. By the time the clone turns twenty and is really able to help the original woman with her kids, the original will be sixty, and her kids may already have their own kids, and in turn be themselves overwhelmed and asking me for clones.

Moreover—and this is the most important difference when it comes to human clones—you cannot copy a person. It's impossible. That clone is going to grow up in a different time, most likely in a different place, with a different diet, a different culture, and different moral codes. People will interact with her in a different way. School will be a completely different experience. Sure, at twenty, she will look a lot like that overworked mother I once met—but she won't *be* that overworked mother. Time to go back again to the leitmotif initiated in chapter 2 and about to be repeated until the end of this book: *We are not a mere product of genetics.* To become a supermodel or warrior, we need a huge amount of training and discipline, together with what we get from our parental genes. We are part genes, part environment, with a strong influence from the stimuli we receive until we are four years old. Why do we keep repeating ourselves on something this obvious? Because we have long learned that it is not obvious. You would think that, by now, all these matters would be settled and clear in everybody's minds, but they are not. Among several other dangerous confusions, these ones still pop up whenever I'm asked to speak to students and teachers at high schools about mammalian cloning. Believing what is described about cloning frame by frame in the film version of *Jurassic Park* is another recurring problem. T-Rex and Dolly came into the world too close for comfort.

SOME PERSISTENT CONFUSIONS

Of course, people cannot be blamed for believing in nonsense if there has hardly been any effort to give the public the correct context and information. In a world where citizens can't even learn about normal sex without a struggle, how can they learn about cloning, a field that involves reproductive technologies and arose from science fiction horror stories? Without clear knowledge of sex and with information provided by science fiction movies, what other context have we for public discussions of therapeutic cloning using stem cells, freezing a baby's umbilical cord stem cells, or freezing oocytes or embryos? Ask anyone not involved in biology research, and, for the most part, they will remember interesting Internet sites—and they can show you where they are—announcing that a clone of Elvis, prepared from a sample of his hair, has been made with success and is now being developed in someone's uterus, or that Jesus is being cloned from cells left in the Shroud of Turin. Right after Dolly, shady doctors went as far as announcing that they were going to clone a dead baby so loved by his parents that they had to copy him, and people still vaguely remember and believe them.

My friends working at veterinary clinics often explode in frustration over the distraught pet owners who show up with a tube filled with the blood of their beloved, just-deceased Fifi: They love her too much; another dog won't do. They want an exact copy, they don't want to hear about all the shortcomings, they just want to know how long it will take and how much it will cost. They're paying for Fifi, period. Most of my friends decline. Some of them are so moved that they try to do it for free—but everyone knows of someone with no scruples who charged a fortune for a hopeless operation. Since an individual dog's behavior depends not only on its genes but also on in utero factors, maternal care, and human encounters, those desperate owners can never get "their" dog back. Even when the vet goes ahead with the procedure, most of the cloned dogs are stillborn or sick. Interestingly, those animals that have survived appear to be quite different from the originals. This is something that the sheep cloners already knew (Wilmut et al. 2000, 5). Even clones made from the same embryonic cells have different "personalities." Why does this happen? We have to be modest: We truly don't understand.

HOW THE SCENE CHANGED

Through all this mayhem, the age of innocence ended with Dolly. Mammalian cloning became a completely different scene, much more competitive and focused on where the next great profits were going to come from. From early on, there were bets on xenoplastic transplants[7] (injecting a nucleus from one species into the egg of another species); on creating different techniques to clone animals (such as pigs, which were of great interest to farmers, as they seemed to resist all cloning efforts); on using cloning to save endangered species by keeping viable cloned embryos in coolers until better conditions were created; on resurrecting some relatively recent ancestors of species living today (such as cloning a mammoth from an elephant egg); and yes, obviously, on stem cell technology. For all the creativity suddenly at work, our meetings acquired a bitter aftertaste, because the vast majority of techniques, and even reagents and culture media, had been patented instead of being reviewed by peers and published in scientific journals. For the first time I could remember, my friends and former officemates were standing in front of their posters at science meetings, guiding me through their ideas, their experiments, and their results, but suddenly coming to a halt, taking a deep breath and saying, "I can't tell you anything else because the rest is the intellectual property of my company." In all honesty, at first I thought they were joking. Quite obviously, they were not. The times, they were a-changing.

STEM CELLS

From the very beginning of the post-Dolly era, there was increasingly intense talk concerning the big jump ahead to a fantastic new type of medicine. It truly would be all gain and no pain, and we would reach it through the use of stem cells. The vast number of rumors starting with "I didn't tell you this, but . . ." that circulated among labs were even more ingenious than the conversations people had out in the open. Stem cell therapy sounded like such an incredible benefit to humankind that it was almost too good to be true. If it were ever to work, it seemed just about miraculous, even to us. It would allow for the regeneration of certain organs in our bodies, just by colonizing them with cloned totipotent cell cultures obtained from the inner cell mass of cloned blastocysts (as mentioned in chapters 5 and 7). Since the inner cell mass, present only during the blastocyst stage, contains all the cells that give rise to the hundreds of different somatic cell types we have in our adult bodies, its cells have to be capable of becoming anything needed throughout our entire development process, be it bone marrow, skin, intestinal walls, nails, lymphocytes, heart valve components, neurons, or any other cell type.

We imagined ourselves as the biomedical researchers of the future, explaining our method to an anxious patient with a failing liver. We chose the liver for perfectly legitimate reasons. We were not assuming that our first patients would all be great geniuses suffering from terminal cirrhosis, although that certainly was a romantic possibility. We were rather sagely picking the liver because it's a simple organ and we needed to start with something simple to try our hand at organ regeneration. Also, owing to its cleansing function, the liver has a great endogenous capacity to regrow its own cells—and inbuilt regrowth programs would certainly help us in our job.

Stem Cell Therapy for Dummies

"Fear nothing, Mr. Jones," we imagined ourselves saying. "Rest assured, we'll fix your liver, and it won't hurt. Let's grab an easy cell from your body. Since we can clone from any adult cell, we'll just grab a cell from your skin, because skin's easy to reach. Now we are transferring this cell to an enucleated egg from a donor. The egg starts developing, and, fine, see, we cloned you. Now your clone goes to the incubator until it becomes a blastocyst with an inner cell mass. At that point, we'll retrieve the inner cell mass from this blastocyst's surrounding cell wall and transfer it to a Petri dish with culture media. After a while, we'll have a culture of your pluripotent cells, the ones we call embryonic stem cells because we got them from the embryo. And so, before you know it, we'll have ourselves a huge culture. Maybe we can even spread it to more Petri dishes, or maybe we can make it immortal. Ideally, a day will come for all these star turns, but enough of our dreams, let's get back to your case, which, by the way, is just about solved. We already have what you came here for. We have a whole lot of cloned stem cells, and we are simply going to inject them into your diseased liver. Once they're there, your own surviving liver cells will do the rest of the task. They'll act upon these pluripotent newcomers, they'll tell them they're inside a liver, and they'll induce them to become liver cells themselves, since, until proven otherwise, undifferentiated cells become what their surrounding cells tell them to become. But, unlike the liver you had before, you will now have a liver made of shiny, beautiful, freshly born-again liver cells, and it didn't take more than two weeks for the whole thing to be completed. That's it."

Quite obviously, we made this process sound awfully simple, and, moreover, we ignored a whole lot of pitfalls and limitations that were not mentioned in our latest grant proposal and that our referees will soon massacre us for. For everything we said above, there are a huge number of "buts" and "ifs" and "hows." There are even furious colleagues right down the hall already screaming that this is terrible science. "But, hey, Mr. Jones, come on: Didn't we give you your liver back? And how does that feel, huh, to be on your own, twenty years younger in no time at all? Please keep your money. Human guinea pigs are on the house. We still refuse to work for profit. History will remember us."[8]

Our First PR Mistake

Sure enough, even while saving Mr. Jones's liver, we were already making our first public relations mistake in stem cell therapy—and this would be a mistake with dramatic consequences. We were telling him that his redeeming stem cells were *embryonic* stem cells. Why did we do this? Well, we were scientists. We care a lot about making sure that our terminology is correct. We were talking about cells retrieved from the inner cell mass of the blastocyst, and the *blastocyst* is an *embryo*—often not yet even nested in the uterine walls, always invisible unless you use a microscope, but scientifically called an *embryo* nevertheless. It also happens to be exactly the embryo with the perfect pluripotent, absolutely plastic cells that are so promising for stem cell therapy. However, we now have induced pluripotent stem cells that we can use for therapy (see chapter 7), and, for those, we don't really need embryos. We can take adult cells from the skin, activate certain genes in these cells, and have the cells revert back to that totally plastic pluripotent state in which they can become all the different cell types of the body. And from here we can continue the procedure as described to Mr. Jones in the most basic terms possible. Therefore, when we first communicated our ideas to the general public and to the media, before we even considered the risks of our bio-jargon, we did speak of, yes, *embryonic stem cells*.

Before a true catastrophe, the Zulus have a fabulous one-word way of expressing themselves: *Tumamina*.

My God, what have we done?

We said *embryo*. People heard *fetus*. Important people in the media heard *fetus*. Decision-makers who should at least seek better information were so horrified when they heard *fetus* that they didn't even bother contacting us (see chapter 1). I remember that our then–prime minister heard *fetus*, and I wrote him an open letter that no paper wanted to publish at first. It was like a weird collective hysteria: Everybody believed, all at once, that we were dismembering those adorable little things from films and pictures, with big bellies and big heads, happily drumming their hands and sucking their thumbs as they comfortably floated inside amniotic fluid, ignoring gravity—it couldn't be happening, but it was. People believed we were killing beautiful new lives about to be born, we were chopping them into small pieces, we were inducing abortions, we were doing things I'm not even capable of repeating in these pages. And they thought we were committing all these atrocities just because we wanted to save Mr. Jones's liver.

Had we used the term *blastocyst stem cells*, probably nothing would have happened. But now we were in dire straits everywhere we went. Trying to get people to tell an *embryo* from a *fetus* was a very difficult job—I lived through a couple of talks where the job even became risky, and many others where the chairperson asked the audience to please not ask me any questions since evil was not welcomed in the room.[9] My university was part of a biology group meant to standardize high school literature on stem cells so that all European students would finish their studies having the same ideas concerning stem cells from the inner cell mass and their origin. This was not controversial until two or three years before the century changed. Then the term "embryonic stem cell" came into use, and European textbooks never unified their language since they couldn't even unify their authors.

Umbilical Cord Stem Cells

Can we stay away from manipulating any embryos of any sort, so that nobody would fear stem cells? It is true that only early embryos have the *pluripotent* stem cells we've been talking about so far. But time marches on, and now we can take adult cells and treat them in ways that make them resemble these pluripotent embryonic stem cells. Moreover, as it turns out, there are also *multipotent* stem cells, cells capable of forming a subset of body parts. These cells can be found in numerous other places, one of which is the umbilical cord. Sound familiar?

The umbilical cord is the lifeline of the fetus to the mother, and it is routinely discarded after birth. But it can be used as a source of multipotent stem cells. Again, we hear claims that such stem cells will allow us to live much longer by rejuvenating our organs as they start failing us, in a simple process that is all gain and no pain. Will this new medical miracle of body repair ever become a reality? In part, certainly. There are already examples of cord stem cells being used in genetically related people (such as siblings) whose immune systems will not reject them.

Indeed, bone marrow transplants are nothing other than the transplantation of multipotent blood stem cells. These multipotent stem cells are capable of becoming red blood cells, white blood cells, and lymphocytes.

Two important facts about cord blood banking are that it can be done relatively cheaply and that umbilical cords and cord blood are relatively easy to obtain. It doesn't even hurt. This means that just about any hospital could, in theory, bank a person's cord blood as soon as he or she is born. Indeed, there are already counties and municipalities that are actively collecting cord blood for public banking. Amazingly, though, most people are not aware of this option. Equally amazing is that it is only available in some places. In the New York City area, for instance, only those babies born within the city limits are eligible. Children born in the suburbs or anywhere else in the state do not have this option (NYSDH 2013; NYBC 2016)

In the absence of widespread public banking, the private sector has seen an opportunity. You, too, can bank your child's cord blood . . . at a cost (box 8.1).

BOX 8.1: "IT'S ALL LEGAL"

The cynic's way of explaining why biotech moved ahead with umbilical cord stem cell banks would be to reduce all the motivations to money and ignore other potential motivating factors. At times, it was hard to stay quiet when private companies mushroomed all around us, banking something that we are not really sure will ever be any of use.[10] They tell us that to be a good parent, we must pay a lot of money to harvest the stem cells of our baby's umbilical cord at birth so that they can be frozen right away and kept in a freezer for later use, when the grown-up child needs them and when the technology will surely exist. It's true; there are several ongoing studies looking at how to make this technology work, and the results so far are encouraging. But the technology for replacement therapy using umbilical stem cells really isn't here yet, and we don't know when it will be. For all the good intentions that could have been at play when this offer was created some ten years ago, we were always facing a familiar and complex problem.

This was often applied science from the get-go, and applied science is usually done for a profit. A number of perfectly legitimate questions could thus be raised—especially when the people involved didn't seem to care about stem cells at all, and furthermore believed they didn't need to in order to sell their services to gullible parents. Since this was simple business, anyone with financial support and an entrepreneurial spirit could easily jump into the arena. This is not necessarily always bad, but it can lead to eerie true stories such as the one that follows, from when the science was in its infancy.

At around 6 PM at my university in Lisbon in January 2003, some ten minutes after I had returned from a two-hour marathon lecture, a peroxide blonde with bright lipstick and heavy makeup entered our office sporting a matching pink-and-blue corporate suit.

"Hi, I heard you're an embryologist."

"Well, yes, this is the developmental biology program."

"And I heard you're also a journalist."

"I have been . . ."—*handshake*—

"I'm a jurist."[11]

"OK?"

"My husband has an MBA and has been a CEO of some companies."

"I see."

"My husband and I just started an umbilical cord company."

"So, you came here because you're hiring biologists?"

"No, we just want to hire you to write for us."

"Write what?"

"You know, to promote our work."

"I see . . . but *who* is doing your work?"

"Oh, two technicians."

"With no supervision?"

"Well, no, the deal is, we have a partnership with an American company. They sent someone over here with their kits and their pamphlets, my daughter is translating the pamphlets, that person trained our two technicians, and now they know how to collect and freeze the cells with the kits."

"And then what?"

"Then we ship the kits to America; it's all legal."

"So you don't think you would need a scientist to follow your work."

"No, really, what for?"

Later on, I saw a couple of famous, young, cheerful-looking morning TV announcers whose pregnancies had been closely followed by the romance press running ads for this company. What can I say? Maybe I'm old-fashioned, but it all sounded awfully scary to me. It's not the clones that scare me. It's not the stem cells that scare me. What's really scary are ordinary people with a lot of power and money.

The danger of improperly regulated stem cell treatments is not a dystopian fantasy. It is all too real. In 2016, three women suffered severe and permanent eye damage when a publicly traded, Florida-based stem cell company injected multipotent stem cells (similar to the ones obtained from umbilical cords) into their eyes. The women had been told that this experimental technique might cure their age-related vision problems. One woman became totally blind and the other two lost much of their vision. Each had paid the company $5,000 for this procedure (Kuriyan et al. 2017; Grady 2017). The patients had also been told that this procedure was being performed by a medical doctor. In fact, the person injecting the cells had no medical degree (Grady 2017). Stem cell biologist Paul Knoepfler (2017) writes that there are about 600 such stem cell clinics in the United States, many "operating generally without FDA approvals, lacking preclinical data to support what they are doing, and experimenting on thousands of patients for profit." In the United States, there is no ban on for-profit experimentation using a patient's own stem cells.

It is indeed possible that stem cells will eventually cure such diseases. But at this moment, such testing is best done at university health centers, where patients do not pay for being in experimental studies and where proper tests are performed so that the procedure is not done if there is a chance it might cause severe harm. You would at least expect that they wouldn't inject both eyes at the same time.

V

EPILOGUES

We end the book with two epilogues. The first is a plea from the heart to the brain. It is a plea for compassion, understanding, and respect for those who are going through or who have "failed" the trials of ARTs. Although affecting nearly 10 percent of the world's couples, infertility remains a taboo subject. It is not discussed by those who are afflicted by it, for whom it is often a source of private shame and pain. Chapter 9 exposes the pain of infertility and the numerous types of damage that this secret pain has caused.

The second epilogue, in chapter 10, is directed from the mind to the heart, presenting a way for the body to be mindful of the improbable wonder one has become. The wonder of the body is celebrated, enhanced by our new knowledge of its origin. In wonder, we see the intertwined strands of curiosity and awe. Fear of the body is also acknowledged. In this, we have both the fear that we will not reproduce a biologically similar child, and the counter-fear, for the earth, that the biosphere is being destroyed by our species' overproliferation.

Both epilogues, in their different ways, concern respect, thankfulness, and both social and individual activism.

9

INFERTILITY WARS

—

How Life Feels After Everything Fails, and, By the Way, How Do We Survive It?

CLARA PINTO-CORREIA

Our true birthplace is the place where one first looks intelligently upon oneself.

—Marguerite Yourcenar, *Memoirs of Hadrian*

As developmental biologists, blessed by the choice of a career that allows us to grow old and become wise without ever having to let go of our childish sense of wonder, we certainly cherish that wonder when we contemplate the making of our bodies. Just like everyone around us, we often shed tears over people whose bodies have been visibly disfigured, deformed, or destroyed by disease, warfare, trauma, and hunger. But, whenever we raise the topic of not being able to have children, we come to learn quickly that not one single tear is allowed—certainly not if we mean to explain what that tear is all about. Because the world around us seems to insist that we're not allowed to cry over being babyless, we learn early on to hide our pain and make it through life as though we could always be tough as nails no matter what. Still, do we suffer less just because we can't suffer in public?

Or does our clandestine condition hurt us even further?

Personally, I didn't have to organize any research effort to notice infertile bodies were not considered worthy of any special consideration, much to the opposite. In all honesty, from the early 2000s onward, as soon as I had collected enough meaningful data on personal suffering from both women and men, I began to lose track of how many times I was told to shut up at ART talks and meetings because what I was saying was irrelevant. Whenever I gave addresses on "forbidden pain," I could tell that the physicians and scientists in the audience quickly grew restless. All those highly qualified professionals with vast hands-on experience had consistently sat down expecting this *pain* in the title of my talk to refer to nothing more than some easily dismissible physical anguish they could easily alleviate, such as the physical pain experienced during embryo transfer that any good local anesthetic would take away. Likewise, the use of *forbidden* was expected to refer to nothing other than substances some people took before surgeries.[1] Technicians, doctors, and biologists listening to me did not want to hear about patients feeling forbidden to shed tears over failed cycles. Let those women go home where they won't bother anyone and cry their hearts out, but please don't come in here bothering us with your annoying collection of personal stories. After all, how was listening to how miserable patients feel before their own failure going to lead to any scientific progress?

Assisted reproductive technologies (ART) are new in the history of human culture, and they are forcing us to look at ourselves differently. At the same time, they are bringing about new experiences we had never encountered before. None of this is easy to deal with, but we haven't even made it to the hardest part yet. Through all the previous chapters, we kept repeating that ART procedures stand many more chances of *failing* than of succeeding. And those failures, in themselves, constitute yet another new experience for all who live through them—this time the experience of a form of violence that human beings are suddenly meeting for the very first time in their history. It is therefore now time to deal directly with that violence, as we turn our full attention to the huge price paid by all those who enter this arena seeking a child they will never bring home at the end of their quest.

There are many questions at stake as we enter these places of last hope. For instance, how seriously do we take the heavy toll imposed on our fellow human beings by these endless possibilities of finally having that biological child they want so badly? Are we thinking enough about the brutality of the experience of serial failed strategies eventually leading nowhere? In consequence of this kind of neglect, and mainly since ARTs were created to make people happy, how come by now so many people are coming out miserable? As developmental biologists and as humans, shouldn't we do our best to address these alarming shortcomings?

Reading through the specialized literature, it is both baffling and demoralizing to confront the number of peer-reviewed papers, chapters, and books overflowing with data on the breathtaking numbers of otherwise healthy people who could have lived happily ever after in many other ways but chose to enter an infertility clinic instead.[2] We soon discover that modern couples often start trying one simple treatment in their twenties and end up tearfully talking with therapists late into their forties, after continuing to reach out to no avail for solutions such as egg donors, intracytoplasmic sperm injection, and foster uteri. This means over *twenty years* of what could have been a productive and happy life of marital togetherness wasted on invasive treatments that led to such recollections as "The raw bodily experiences—being poked and prodded with various sharp implements in the effort to get pregnant, bloating from the medications, checking for blood in underwear or on toilet paper and too often finding it" (Josephs 2005, 33), "Having struggled with issues of loss of control, the existential meaning of infertility, and the feeling that I was 'going crazy'" (to the point of no longer recognizing herself, she later admitted), and "I quit my job in hopes of 'increasing my odds' of pregnancy, felt alienated from family members who were able to have children while I was not, and eventually had the issue of infertility affect my marriage" (Burns 2005, 16).

This is not to mention the infamous "mood swings" most women undergoing hormonal stimulation report, while those around them insist they're just having "psychological issues." Or the eventual cases of women literally developing "needlephobia" (Rosen and Rosen 2005)—which might sound really whiny at first but makes sense if you actually bear in mind that women have to give themselves two hypodermic injections a day for two weeks, each cycle, not counting all the needles needed for egg retrievals and in vitro fertilization (IVF) implantations and other procedures at the clinic. I can relate.

And all of this is going on while the husband first tries to be supportive but almost always ends up feeling useless and unwanted, finally responding to the whole mayhem by both losing his libido *and* watching sports on TV with a terrible frown for hours on end.[3] The responses of men to infertility have not been as widely studied as those of women (see Throsby and Gill 2004; Hannah and Gough 2015), but my experience is that guys seem to obsess considerably more than women over the tired issues of "their" genes and "their" offspring. Without breaking confidence, I can even say that I have heard amazingly similar fantasies from numerous childless married male professionals about running away with the poor-but-fertile stair cleaner whom they lately lust after, perceived as a potential and thankful mother for the offspring of their dreams.

Therapists reporting on couples who "fail" ART[4] often comment that young couples with no children first enter an IVF clinic feeling absolutely healthy, only to rapidly start considering themselves as patients who then act and interact as such. After all, their bodies have "failed" the test. It's generally not considered that the test rather failed the couple. It can only make sense that the latest publications on the consequences of any failed IVF procedure mention the need for mental health professionals with relevant expertise (Sclaff and Braverman 2015) and discuss depression and suicide in unequivocal terms (Vikström 2015).

With this information in the background, none of us needs to be a therapist to imagine why women writing books on their personal encounters with round after round of failed IVF often give their works biblical-accursed titles such as *Give Me Children or Else I Die*, the desperate plea Rachel makes to Jacob when she sees she "bore him no children" and becomes jealous of her sister (Genesis 30:1).

INFERTILITY VOICES IN MODERN LITERATURE

There is now a wealth of literature created by women writing books in this vein. Many don't spare us one single detail of what they have gone through, as though their experience had been such a heavy weight that they had to share it all in order to get it off their chests. Or else they're reporting that their experience totally changed them, they decided never to have children, and they are now researching the "hidden population" of the voluntarily childless (Wilson 2014). They might also tell us that they remained childless against their will and now are collecting testimonies from other women on how they coped with the same situation (TheNotMom.com 2016). Or, they learned about the secret pain of infertility through a childless best friend and were so impressed that they dedicated an entire volume to childlessness in the United States (Vissing 2002). In turn, this literature spawned a rich subvariety of other related stories. This time we're talking about a true all-you-can-eat buffet: a close look at the real reasons an increasing number of men choose to remain childless, often involving job security (Lunneborg 2002); a biological mother going to the movies with the surrogate (Bialosky and Schulman 1999); lesbian couples becoming edgy because they keep failing at repeated cycles of artificial insemination with sperm from anonymous donor (Goldberg 2010); gay couples meeting their surrogate and her husband, wondering whether straight couples would ever be put through such tight scrutiny, and later a partner becoming jealous of the other partner for having the "best" sperm (Peterson 2003); and even biological mothers who gave their babies up for adoption and now want to meet them as twenty-year-olds but keep hearing horror stories about the outcomes of such reunions.

These oft-ignored autobiographical authors are not exactly bored suburban housewives in bad need of a child just to fill their otherwise empty days. For the most part, the women who publish books on their struggle through the infertility wars are strong-willed professionals such as writers, lawyers, doctors, and professors, some openly feminist (Alden 1996), many referring to their partners as lovers or companions rather than husbands (Wilson 2014). Their first-person-account books form a genre of their own, dating back to the early 1990s, as soon as IVF became a familiar enough technique for couples to expect it to work immediately after going through a few years of failed attempts to achieve pregnancy on their own. From the start, these books have included a vast spectrum of tales and tastes that can be as soapy as that of the self-described "Jewish princess" in *Full Circle* (Diamond 1994) or as survivalist as that of Proud Mary in *Crossing the Moon* (Alden 1996). They can finish with the pained acceptance of life without children, or rather with the deliberate choice to become quite militant about the right to be childless, or even with the development of a mutual support network born from the harshness of the experience. My own book (Águeda-Marujo and Pinto-Correia 2004) was based on long interviews between me and my friend, a specialized psychologist volunteering at our infertility helpline. My friend told me that I was surviving my pain by laughing in the face of sadness and constructing horribly raw metaphors to describe it. She still won't dare try to publish them.

COMMON TRAITS OF FIRST-PERSON ACCOUNTS OF INFERTILITY

Most books of personal infertility accounts share an impressive number of traits. Titles tend to be bleak, and book covers tend to be instinctive downers. Their leading ladies frequently start the IVF journey with a man who stands by their side, and then, little by little, as hope slides away and the road ahead becomes harder and bleaker, that man just about vanishes from sight. The authors' choices of metaphor for the growing emotional and physical pain that develops as the IVF war unfolds blows up to bolder and wilder proportions. At some point, they grow seriously aloof. There always comes a time when they can't come anywhere near a woman who is pregnant or holding a baby, even if that woman is their own beloved sister. Giving up is almost unspeakable—many authors just skip that moment, like heroin addicts tending to skip the moment when they stop using in their own autobiographies. And there is yet another extremely interesting detail, worthy of further attention. At a certain point, it becomes impossible not to notice that our unlikely heroines increasingly suffer because they never qualify for social sympathy, unlike those with a painful handicap or history—say, the blind, war veterans, the homeless, or refugees.

Now, those who were blessed enough not to have been drafted into the infertility wars may not understand this particular problem, but anyone who has been there knows how bitterly it stings. Just take this simple slice of life to start with: Most people give up their seat on the bus to a pregnant lady, but nobody gives up their seat to a woman overdosing on progesterone to increase the odds of implantation after embryo transfer. Yet, their symptoms are quite similar—except the first one is blissfully anticipating happiness in public view, whereas the second one is anxiously wondering what to expect in total secrecy. This is hard enough to live through, but it becomes even harder as you learn from experience that your total secrecy should definitely remain secret, for the sake of your own peace of mind, and often for that of your loved ones, as well. I'm both fond and weary of my real-life anecdote of the family therapist, since it is a perfect illustration of this tough way of learning a lesson.[5]

It happened at a dinner party at my house in Massachusetts one gorgeous winter night, featuring great food I had cooked with a lot of loving care, with everybody sitting around a lovely table likewise set by myself with all the attention in the world paid to the minutest of details. The family therapist showed up as a friend of friends and proceeded to make everybody laugh with tales of his chemical experimentations in college. Everybody seemed to enjoy both my cooking and my decorations, the wine was flowing nicely, and we were all having a good time. All went well until, for some unfathomable reason, perhaps just because I had drunk enough to let my guard down, I dared to mention my then-ongoing struggle with infertility. The trigger might have been the fact that I had uttered one of the most secret of our warfare words: *pain*. The family therapist nearly jumped from his chair in anger: "What do you mean, *pain*? I was drafted into the Vietnam War and had to watch my friends die!" And that was just about the end of the enjoyable dinner party. Everyone did their best to hurry to their coats and their cars and disappear through the white veils of snow.

Yes, of course, we all have our pains. Many of those pains might be worse than infertility. Having spent my childhood in a war zone in Africa, I admit that it is worse to be drafted. Having heard stories told at the military hospital in Luanda, during the Angolan civil war, I would be the first to agree that it is worse to watch your friends die in combat or be disfigured forever. Having been blessed with excellent eyesight, I guess I should thank God to have afflicted me with infertility rather than blindness. Having seen fertile friends with severely handicapped children,[6] I will promptly add that I personally would prefer not to have any children at all than to fight that other kind of devastating battle. Having watched my grandfather, then my aunt, then my father, die from cancer after several

decades of splendid health, I have often joked that at least my own share
of life's unfairness would not unexpectedly kill me in that most ignomini-
ous way. But here I was, and I had cleaned and cooked and prepared that
dinner party with such heights of neurotic attention to detail only for one
simple haunting reason: I had become an obsessive perfectionist because
I just didn't want to have a single second vacant in my mind by day or by
night, or else that microscopic empty space would immediately be filled
by the fact that I was never going to have my own children. Then my pain
would once more proceed to become monstrous enough to kill me all
over again, since at that point I felt like I had already died a good number
of times, each more unpleasantly than the last. I had always loved to have
our friends over and spoil them rotten; but now I was frankly overdoing
it, just so I had no time to think and die again. Then this man comes along
and tells me my pain is preposterous, no one bothers to side with me, and
everybody hurries to get out of the storm.

This pattern hasn't changed a bit since my own infertility took away my
childhood dream of building a huge and happy family around me, and
there are obvious social reasons why.

We all feel strong sympathy for the drafted, the blind, the parents of
children with intellectual disabilities, the victims of cancer—anything
plain to see. However, when it comes to infertility, much as simple endometriosis[7] can physically hurt to the point at which a woman can't walk,
we still are not even allowed to use the word "pain"—not even before
family therapists, during a party thrown by ourselves and held at home
within a circle of friends. You have to respect the power of a taboo when
you meet one.

And a taboo is a heavy load to carry.

Many couples dealing with all these levels of distress often end up feeling locked inside a cage that nobody else can see. They start to cross the
warzone holding their breath, waiting to exhale. They develop an increasing anxiety over the next possible landmine hiding under their path. Over
the years, several women have told me that, past a certain point, they
started to feel that even their own doctors lost patience at continuing
to hear about their anxieties and sorrows. For as long as they insist on
remaining in the chase, these couples' marriages drift further and further
apart from anyone's idea of marital bliss. You have no idea how stories of

happily ever after can go wrong until you're listening to women telling it all at infertility helplines.

A RICH LITERATURE WITH AN ABSENT READERSHIP

Who cares about these authors of first-person accounts of infertility? Who's reading them? Yes, these authors certainly have given us an interesting wide range of a brand-new literature awaiting further studies. But that said, who's their readership? Here is the other notorious piece of information that can be readily checked by anyone interested, just by looking at the records of a decent number of libraries:[8] Almost no one reads this literary new genre so rich with new insights on the human psyche, where life's existential meaning is so intensely documented. With so much information on ART out there, including juicy scandals and the indescribable stuff reality shows are made of,[9] you would think that, by now, directors would be using this literature to make Hollywood movies, TV series, or even Comedy Central recordings. Or, on a brighter note, there could be PBS documentaries. Other people could study these books to write award-winning novels or to prepare really interesting NPR talk shows.

There is even something to be said for the documentary and fantasy value of self-published romances like *The Baby Game*. Here, a middle-class couple from Arizona (not to be confused with the desperately childless couple in the fictitious film *Raising Arizona*), for whom, as usual, adoption seems to have been ruled out by default, decides to go to India to inseminate a surrogate. This other person is presented as a nice family-oriented woman, who benefits greatly from the deal, using the money from her surrogacy to alleviate her miserable life in a Bombay slum. Eventually, the couple brings home their adorable half-Indian twins, and they make their final flowery utterance of the mandatory "I just want to share my story to give hope to others." Procedures of this sort are so highly debatable that by 2013, the United States government forbade Americans to procure themselves surrogate mothers in India (DasGupta and Dasgupta 2014). By then, India had already earned the dubious reputation of being "the world's baby factory," "the reproductive assembly line," the "mother destination" for commercial surrogates, and the country where "giving birth is outsourced" (Inhorn 2015). But none of this seems to have bothered either the well-meaning Arizona couple or the author of the book on their success story—not to mention their agent or their publisher.

First-person accounts need not be taboo or embarrassing. The pain brought about by life with infertility just needs to be open. The wealth of literature available has the advantage of providing all those interested with easily understandable tales that can lead them to much-needed sources of information ready to use. Today, college students could be reading personal stories from the infertility wars as a twenty-first century literary spin-off of biomed technology. It is a new literature and will undoubtedly keep evolving, which could offer academia new venues of literary research. Or, quite simply, individuals could be reading these new books for the sake of curiosity, education, empathy, personal growth, and even self-help. However, when we verify their circulation dates, the patterns are dismal: Most of these books do not even get to leave the library. And, to top it off, some of them are not even placed on the shelves: They are kept in depositories, those hidden bowels of libraries where staff keep the books nobody ever asks for. This seems to be a universal pattern, and it is one I have experienced personally: For all the promotion efforts of my publisher, our infertility book hardly sold at all. We never got to start the public debate that we had hoped we would, and therefore the improvements in the infertility helpline office that we had planned to make with the money resulting from our royalties had to come instead from our imagination. Only people calling us after a visit to an IVF clinic mentioned having read our work. Couples in the infertility wars are true loners like few others. As a rule, they know it. If they don't know how alone they are, they soon learn at their own expense, taking their silence home with them, exposing themselves to even more serious damage. But maybe they don't know it yet. And who's there to tell them, anyway?

The library of unread resources is growing larger and larger. There is an entire Listserv from the Center for Genetics and Society on ART, and there are numerous articles on personal experiences in journals of clinical psychology, feminist psychology, anthropology, and sociology. Most recently, two films on the topic of infertility premiered at the Tribeca Film Festival, and a video exposé came out in Australia. Journals and magazines such as *Wired* are beginning to carry this material and question the cheerleaders of new technologies. The September 4, 2016, *New York Times* book review had a front-page article reviewing two books on infertility (Cusk 2016). One book (Leigh 2016) is another tale of a life and marriage being destroyed by the single-minded pursuit of "our child." The second (Boggs 2016) talks about the desire for a biological child growing stronger with each passing year, instead of deepening into wise acceptance. The author quotes Virginia Woolf's diary allusions to her own infertility: "Let me watch the wave rise. I watch. Vanessa. Children. Failure. Yes. Failure. Failure. The wave rises." Such is the real existential anguish of our technologically brilliant but emotionally silent times. As personal stories finally pile up, maybe they will soon reach a size that makes them impossible to ignore, and then maybe there will be a new readership besides those engaged in the battle. Or not. As blossoming ART keep further exposing us to increasingly complex infertility wars, it remains to be seen whether we are finally ready to deal with a kind of perturbing pain that remained unacknowledged and unaccepted for thousands of years.

COUNTERINTUITIVE LIVING AND COLLATERAL DAMAGE

Everything considered so far might be tough on us, but then stupid little details avidly contribute to make matters worse. Childless couples crossing the infertility war zone have no choice but to dig themselves into a terribly counterintuitive situation. Our background becomes more and more frenzied as the present century unfolds and reduces our attention span to almost nil.[10] We live in the days of attention deficit disorder, hyperactivity, baffling speed, short-hand text messaging, numerous monitors simultaneously flashing information at us, social networks often featuring highly distorted and fragmented news posted and checked in a hurry between other urgent tasks. So here we are, within a greedy society trapped in speed. But then you enter a month-long cycle in a blind roll of the dice. Now you have no choice but to go slow. Those estrogens have to be injected for two weeks twice a day at the right time, and don't you dare not keep them in that cooler at the right temperature. That insulin syringe has to be filled to the exact measure with patient precision, and don't you dare go too fast and allow bubbles to form. That needle has to get *under your skin,* not *inside your muscle*, and don't you dare rush while pressing the skin fold or you'll get a blister and those fantastic and fantastically expensive molecules will get lost. No one is telling you that it is forbidden to continue with your usual frantic pace of life while following all these unusual routines, but then—more often than not—when you don't succeed, you're going to feel really guilty because maybe you caused the failure by moving too much or being too stressed. On the other hand, if you choose to quit your exciting yet extremely demanding line of work in order to dedicate yourself to "a quiet life," chances are that the heavily silent eventlessness surrounding your new sweet self will drive you out of your mind, and might—just might—severely affect your chances, because you ended up growing more anxious and anguished by going low-profile.

You guessed it.

Yes, these endless attempts are generally combined with periods of profound clinical depression, for one partner or both. Yes, they often destroy marriages. If one of the partners involved rapidly proceeds to remarry somebody younger and soon afterward produces a parade of children, the previous partner is seriously hurt again. So yes, of course, infertility wars can destroy lives.

And yes, this has all happened many times before.

It is not that one of the partners was mean or heartless, or that the woman trying to get pregnant made a childishly misguided choice when she quit her great job for the sake of a "quiet life," which turned out to be full of inner turmoil, while trying to increase her odds of getting pregnant. It's just that we are all human with our foibles and wild sides, and our deepest darkness is too easily exposed by the overarching demands of an IVF war with no end in sight.

Until all those surrounding us understand involuntary childlessness as the extreme pain that it is and has always been, for as long as there is a historical record of human behavior,[11] we will have a serious problem that profoundly damages many, many people. Suffering the extreme pain of involuntary childlessness has been worsened by ART, and this new literature documents it.

Having dealt so far with what this new literature tells us, we should now also pay attention to what it makes a point of not telling us. Some passages of our personal journeys an easily be held in low esteem due to their concessions to poor taste. And so, instinctively, we leave them out. But should we?

People who write about their own infertility dramas focus on their childlessness, and they tend to skip anything that doesn't relate directly to each failed IVF cycle. As my friend and I did in our own book concerning my personal experience, other authors prefer to leave out their ways of seeking any sort of pleasure while making it through hell. Most of all, they skip any account of how they lived sexual lives when they were trying to get pregnant. We can't blame them. Most likely, we should thank them. Too much information can be a terrible thing, and it tends to damage our best projects. But these elegant ellipses create yet another problem.

Other than through those books, infertile people hardly talk to each other. Sometimes they indulge a little bit here and there in waiting rooms of clinics or hospitals. Sometimes they seek meetings organized by helplines such as mine. Sometimes through those meetings they make real friends, with whom they talk a lot. However, even when we talk of our infertility treatments in gory detail, we still prefer to skip the loaded sexual details of our daily survival. There is nothing all that wrong about this silence. But, for as long as this silence prevails, each infertile woman seeking solace out of sight will think she's the only one crashing that low under her own stress and torture, and she will be terribly ashamed of herself.

It is important to state publicly, once and for all, that it is truly not uncommon for people put through the ART treadmill to eventually deviate from social norms because they just cannot take it anymore. Once and for all, none of us is alone in this. None of us. Again, let he who is without sin cast the first stone.

Over the decades, without ever passing judgment but just holding many hands and listening to many stories just like mine, I've seen men and women put through IVF cycles doing it all. They slide into incredibly complicated affairs halfway through the process for the sheer need of the kind of relief only the comfort of strangers can bring. They have elaborate funerals for the souls of their lost embryos years after the fact, because they are still not at peace with their loss, only to divorce a little later because they are still not at peace with anything after all. They get pregnant by somebody else and make believe that the IVF worked just to stop the nonsense—but then the third party decides to have a say in the matter and there goes the neighborhood. This is not storytelling. This is just how fragile people really are when confronted with having or not having their own children, and what it can take to have them. Sometimes it takes too much. We may survive, but not without scars.

COSTS

Next comes the other serious problem that is always at stake: Compounding the physical and emotional drain, there's the fundamental issue of how to pay for expensive ART treatments. Since success doesn't usually come with the first cycle, those who want to keep trying have to keep paying more and more. How do you spend these fortunes and still go ahead with your life? Here is one couple's story:

> As they did after each failed attempt, the couple would retreat to their home and careers, regroup their energy and finances, and then renew their search for a fertility treatment. Years and years of her life with Harry just flew by in a flurry of bills and treatments.[12]

Needless to say, most people can't even start to afford such costs. One cycle is expensive enough. Several cycles, as are generally needed, are terribly expensive. Although there are publicly supported ART treatments scattered throughout the world, the amount of investment governments are willing to put into them varies considerably, depending on the economic structure of the country and on the country's perceived need for having more children to enlarge dwindling populations (box 9.1).

BOX 9.1: PATTERNS IN ART COSTS

The Scandinavians rank high on the ART-support list. In Sweden, all ART attempts are fully paid for up to six cycles. Denmark gets the highest mark when it comes to babies actually being born through ART, a boom that in 2012 already totaled 4 to 5 percent of the country's entire population.

Israel ranks highest in ART financing and legislation: Everything needed is paid for at all infertility clinics, at least twice per couple, not just until a cycle in completed, but rather until a child is actually born (Frenkel 2001). It is also the only country in the world with legal surrogate motherhood for which it is mandatory to keep systematic records of each procedure, and the success rate seems to be the highest in the world. Israeli legislation on this front is quite strict and heavily enforced. For instance, close relatives can't double as surrogates, so as not to create complex situations further down the road. Also, only women who already have children are allowed to be surrogates. These women are well paid, and the contracts they have to sign before starting the procedure have been carefully worked out by lawyers and ethicists, although they are still continuously under debate. Couples who order a surrogacy have to keep the resulting baby even if he or she happens to be born with any sort of physical problem or intellectual disabilityfrom all over the world and highly trained international doctors. You can choose your own program and check out our prices in our enticing online catalogue. Leave it all to us, because even the plane trip is included in the package. And the beaches are awesome, bathed by the warm and restoring Indian Ocean.

Currently, India offers the most reproductive hotel-clinics advertised online,[16] closely followed by Thailand and China, which now offers stem cell therapy (Inhorn 2015). However, the main contemporary "reprohub" seems to be located in the absurdly rich and equally beach-blessed Dubai. Sagely positioned near the coast of the Persian Gulf in Dubai, the *Conceive* clinic caters to infertile couples from five continents and nearly one-third of the world's nations. Welcome to an immaculate setting where Muslim patients from Pakistan seek Hindu physicians from India, are cared for by Catholic nuns from the Philippines, have their embryos handled by Greek Orthodox embryologists, and receive follow-up instructions from two African clinicians, one from Sudan and the other from Somalia. At the entrance, infertile Arabs, Asians, Europeans, and Africans await their turn to tell someone on this team the long story that has brought them here (Inhorn 2015, 17). In 2014, *Conceive* didn't even have a website, but patients already knew about it through word of mouth. According to Muslim rules, some types of ART are not offered there, including surrogacy and egg donation, but this doesn't seem to stop people from coming.

A young Muslim woman from Somalia, living in London and married to a Muslim Ethiopian with an Egyptian mother, brought with her a three-year-old daughter from a previous IVF procedure and two years of credit card debt. She came to *Conceive* because neither her family or her husband's would stop pestering her husband until she bore him a *son*. They would often tell him to leave her and get himself some decent wife, and his mother was constantly calling the house to ask whether there was anything on the way yet. Having tried and failed an absurd number of infertility treatments and cycles in London, and knowing that new doctors at new places were but "a gamble," she explains (Inhorn 2015, xvii), this woman summarized her attempt to get it over with at *Conceive* in just three words: "Society bullies you."

BETTER DAYS AND INFERTILITY STUDIES

Not all IVF patients risk being abandoned by their partners for lack of producing children or have child-obsessed families constantly harassing them. Not all women with blocked tubes who have already had a daughter are forced to keep pushing their limits because their culture mandates that they bear their husbands a son. Not all of us have felt desperate enough to become ART globetrotters. But everybody knows vaguely about such stories, and about many others, because once we're drafted into the infertility wars we have no choice but to hear endless rumors through the grapevine. Even when we hold a degree in the field, like I do, it is not at all easy to separate fact from fiction and divide our ideas into good and bad, if good and bad ever apply at all. As we lose one more infertility battle we feel ambivalent about fighting yet another, and this ambivalence is one more driving force pushing everybody off balance. We all wish our moral guidelines would be clearer, and we all remain stuck in an unclear territory of uncharted emotions. We have all heard of Malthusian curves and of the big population surplus scare that comes with it. We all know the planet is already way too crowded; we all know human reproduction has to slow down before the rainforest is cut down for good. Therefore, we all feel somewhat pathetic, if not downright guilty, for our repeated efforts to bring into the world one more tenant—which in turn makes us feel really stupid, and none of this helps any. Anyone who reads a newspaper or listens to the radio knows that we are overpopulating the planet with our species and making it unliveable for other species. It's even in biology textbooks. If people around us all seem to agree we have to curb unruled human expansion, why do they act as though having children could be considered the baseline for social morality, and why does being infertile remain akin to indulging in some sort of social capital sin?

Once more, with feeling: On average, one couple out of fifteen on the entire planet is unable to have children. One out of fifteen might be a minority, but in absolute numbers, this minority represents a huge crowd. And this crowd is still being instinctively punished, not just because it has always been so, but also because society has now come to believe that there is no miracle baby that ART cannot provide. Life is good. There is no reason to keep on making it bad for a very significant slice of the world's population. There are already too many horrible things going on in the world that we feel powerless about and unable to change. This one, at least, is a front where we can finally make a truce. Looking infertility straight in the eye and talking about it openly could just as well become one of the major sources of relief for the entire twenty-first century. All it takes is finally calling a spade a spade. Among other things, it would force doctors and practitioners to all be equally honest about what they can really promise their patients (CDC 2016; Resolve 2016).

People always need to be told the following:

- Approximately 85 to 90 percent of infertility cases can be treated with drug therapy or surgical procedures. Fewer than 3 percent need ART such as IVF.
- A healthy young woman has only about a one-in-three chance of succeeding on each cycle of IVF. A thirty-six-year-old woman has about one-in-six odds of success.
- The cost of an IVF cycle is around $12,000 to 15,000 per cycle, and many couples end up paying over $100,000 to achieve a pregnancy. The Affordable Care Act of 2010 does not require coverage for infertility treatments.
- A woman will probably not be able to achieve a pregnancy from eggs retrieved from twenty years earlier.

Let's start by showing ART patients some respect and addressing them with ordinary decency. For instance, the picture published in infertility journals featuring a paternal-looking bespectacled doctor in a clinic background addressing a young, anxious yuppie couple under the caption, "You're their only hope," comes across as a terrible idea in very bad taste. Considering the abundance of infertility conditions worldwide, the pain of those affected, and the lack of resources in most of the world to alleviate the plight of the women who potentially suffer the most, together with our moral obligation to succor them all, here is a suggestion: Universal access to reasonable ART could receive significant and efficient financing from well-meaning philanthropic organizations. At 6 percent of the population, infertility is a normal part of the human condition. So let the history of infertility, in all its bitter and exciting detail, come out into the open, so that our kids will know what to do with it when they experience it. It has worked for those advocating for the rights of people with disabilities. It has also worked for people advocating for the rights of lesbian, gay, bisexual, trans, and queer individuals, now incorporated under the respectful LGBTQ umbrella. In both cases, these groups went public with their existence and brought their personal grievances along. Still, it has not worked for abortion rights, just as it has not worked for those suffering from infertility. What do both groups hold in common? In both cases, we tend to keep our pain secret and to feel it as shame.

What on earth should we do next?

Maybe we seriously need to improve our efforts toprovide and access better information on matters that are crucial for all of us, and to do our very best to reason about them without preconceived refusals, denials, boundaries, or shames.

The world is still a beautiful place, life is still a wonderful miracle, and there is still a lot of room left for all of us to enjoy it together. We just happened to be here when ART suddenly took the entire world by storm and started changing at an almost unbearable pace our most sacred notions of sex, reproduction, family, and genetic inheritance. This book was our joint effort to clear infertility from innocent mistakes and voluntary lies, so that at least future readers understand from the start what is it that they are facing, now that birds and bees are gone and there is no turning back. We know it is a modest contribution. In good faith, it is what we can contribute without indulging in opinionated statements our academic backgrounds don't give us any special right to deliver. Although I tried my luck at repeated IVFs, I lack biological children. However, I adopted my kids I soon afterwards, and then promptly caught myself wondering what the big deal had been all about. There are therefore several reasons why I don't think I have to die without ever revealing what infertility made me go through. One doesn't have to have children to influence the next generation. If somebody, somewhere, starts a well-meaning and well-informed discussion on ART and infertility services because of our combined efforts, Scott and I will have earned our day.

10

THE HUMAN CONDITION OF FEAR AND WONDER

———

In Celebration of Bodies

SCOTT GILBERT

As long as you remain under the domination of the delusions and their underlying states of ignorance, you have no possibility of achieving genuine, lasting happiness.

—Tenzin Gyatso, Fourteenth Dalai Lama, *The World of Tibetan Buddhism*

THE CENTRALITY OF WONDER

"I am fearfully and wonderfully made," says the psalmist. But what is it to be fearfully and wonderfully made? How does one respond to the amazement of one's body? I want to go back to the source of these questions: wonder. I will be proposing some hypotheses concerning embryos, wonder, and the relationship between science and religion (see Gilbert 2013).

I profess embryology, the science of how our bodies are made, a science that seeks answers to ancient questions: How did I come into being? How does sexual union generate a new life? How do I come to look like my parents? How come I have only two eyes, and they are both in my head and nowhere else? How do my muscles become connected to my bones? How come some people have penises and others can have babies?

Embryology is a profession in which wonder remains an operative category. French embryologist Jean Rostand (1962) said it very well when he wrote, "What a profession this is—this daily inhalation of wonder." As an embryologist, I'm privileged to experience wonder daily and to *expect* to be amazed when I enter the laboratory. For many of us, though, wonder has become something we experience only on vacations or as a surprise.

I would contend that wonder is a primary experience, the result of the mind encountering the universe. But only mystics, perhaps, can live in a state of perpetual wonder. For most of us, wonder has a short half-life and rapidly decays into two lesser, but still powerful components: awe and curiosity. This is clearly seen in language, where wonder has both these meanings. Curiosity is seen in the English expression, "I wonder." Awe is seen in our declarations of "the wonder of the world." Awe and curiosity both originate from wonder. From curiosity comes the quest for truth about the physical universe and the testing of ideas against other ideas and against experience; that is to say, the foundations of philosophy and science. From awe come the reverence and gratitude that are characteristic of the religious attitude. Science and religion, let me hypothesize, both descend from wonder.

Plato and Aristotle agreed that wonder is the beginning of knowledge. Echoing Plato, Aristotle (350 BCE) notes, "For it is owing to their *wonder* that men both now begin and at first began to philosophize." At the beginnings of modern science, Francis Bacon (1605) reaffirmed that wonder was "the seed of knowledge." Statements of wonder are not uncommon in the autobiographies of our contemporary embryologists, and they are sometimes present even in our scientific papers.

One of the most important statements of wonder in embryology comes from the medieval rabbi and physician Maimonides. He writes (1190),

> A pious man of my time would say that an angel of God had to enter the womb of a pregnant woman to mold the organs of the fetus. . . . This would constitute a miracle. But how much *more* of a miracle would it be if God had so empowered matter to be able to create the organs of a fetus without having to employ an angel for each pregnancy?

Indeed, my job, my career, is to discover some of the ways by which ordinary matter (whether divinely empowered or not) can form itself into an organized embryo. It's amazing. Biologist and poet Miroslav Holub (1990, 38) claims,

> Between the fifth and tenth days the lump of stem cells differentiates into the overall building plan of the embryo and its organs. It is like a lump of iron turning into the space shuttle. In fact, it is the profoundest wonder we can imagine and accept, and at the same time so usual that we have to force ourselves to wonder about the wondrousness of this wonder.

So wonder can give rise to curiosity, which promotes the theorizing and testing that are science. Wonder can give rise to knowledge.

But knowledge is not wisdom. Moses, Jesus, Siddhartha, Confucius, and Muhammad did not know the number of protons in a carbon atom or the four bases of DNA. Knowledge is critically important, but it cannot pass for wisdom. Wisdom is how to use one's knowledge to interact with others in healthy and mutually supportive ways. "Awareness of the divine," writes religious philosopher Abraham Joshua Heschel (1954, 44–45), "begins in *wonder*." For wonder generates not only knowledge, but also wisdom. Heschel continues, "The beginning of awe is wonder and the beginning of wisdom is awe. . . . Knowledge is fostered by curiosity; wisdom is fostered by awe."

Thus, one can affirm the following lineages from wonder: Wonder gives rise to curiosity and awe. Curiosity gives rise to science and philosophy; awe gives rise to reverence and religion. Science and religion are the grandchildren of wonder.

So why should science and religion be fighting against each other? The reasons are mainly historical. In Europe, religion claimed the right to be the literal, scientific truth. Science was done under the umbrella of religion, and indeed, science originated in the West as an attempt to show that a person could have a rational belief in God. In fact, the more detailed the science, the more glory to the creator of such marvels. The intricacies of a bird's feather and the muscles of the hand could be studied in all their details, because such knowledge would give one an even greater appreciation of the Creator. Newton was primarily a theologian, and Darwin's degree was in theology. With such a view, nature was God's creation. As such, it could be taught only by ministers of religion. People like Thomas Huxley and Ernst Haeckel, who wanted to teach biology independently of a religious context, were not allowed to do so. So they used evolution as a way of separating nature from Creation (Barbour 1971, Desmond 1997).

And religion did not at first realize that by claiming to be the scientific, literal truth, it was going down a path that would make it smaller and smaller and less and less relevant. Science became an enemy because it kept providing evidence that the literal reading of the Bible (the Bible as a science text) was wrong. So, as science showed that the world was over four billion years old, and that all the world's species were not created together, religion became smaller and less relevant. God became the "god of the gaps," the god that could only live within what science did not know. Western religion was claiming that what was written to produce awe and reverence in a preliterate Jewish community was scientifically true.

But if the Bible and other religious traditions are not sure sources of scientific knowledge, they remain sources of wisdom. Let me give an example. To a biologist, few stories in the Bible are as silly as that of Noah. Here, two of each (or seven, depending on the chapter of the Bible) of the world's 750,000 known species of insects, 850 species of bats, and every species of worm, salamander, and bird were present on the ark. A couple of ants from the Central American rainforest would be expected to travel across the ocean to the Near East and bring their specific food plant with them. It is foolish to think of this story as scientific fact. But what about taking it as wisdom? Here, there are some interesting items. For instance, one commentary suggests that Noah was called righteous because he went out of his way to painstakingly acquire detailed knowledge of the habits and feeding schedules of the animals so that he could house them properly on the ark (Zornberg 1995). Here we have the beginnings of the notion that one has to know what is true in order to do what is good. Agnostic evolutionist Thomas Huxley (1870) made this an explicit principle: "Learn what is true in order to do what is right." This is one of the reasons we give for learning about ecology in order to save the environment. One has to know the facts in order to not do damage. It is one of the reasons for learning about the body—so that we have the facts that enable us to make effective medicines. Moreover, we can learn from this story that the world can suffer for human moral failings, and that humans have a stake in creation. Now the story of Noah's ark is a parable worth knowing for our times—not for its facts, but for its wisdom.

Yes, both can be right, as long as science does not profess to have moral answers and religion does not profess to hold scientific truths. So, as Stephen J. Gould (1999) notes, there are two "magisterial" ways of appreciating the world: scientific knowledge and religious wisdom, the two grandchildren of wonder. The two must interact.

Science and religion must interact because they both depend on wonder for their existence. Without wonder, science will perish. It will become strictly a means by which some people acquire wealth and power. Without wonder, religion will perish, too. It will become merely a means to keep a subdued population content while some small number achieve wealth and power. If wonder is the source of both science and religion, it is in their mutual interest to form alliances to protect, preserve, and expand sources of wonder. And the major source of wonder remains nature.

FEAR

Thus, chief among the reasons for science and religion to forge an alliance is what biologists call "the preservation of biodiversity" and what theologically oriented people call "the stewardship of Creation."

Our stewardship of this planet has not been adequate. Within the next century, we can expect to witness the extinction of half the animal species on the planet. The human population exploded 2.5 times from 1950 to 2008, reaching a total of more than seven billion people, with many living in desperately unsustainable extremes of excess and of deprivation. Feeding and housing this rapidly expanding population is causing a period of cataclysmic extinction, perhaps similar in scope to the event that destroyed the dinosaurs. It is estimated that one out of every six species will soon go extinct if we follow our "business-as-usual" trajectory of emitting carbon dioxide into the atmosphere (Kolbert, 2014). The increase in temperature we are experiencing comes primarily from human technology, as the amount of resources consumed increased 800 percent during the twentieth century. Wild populations of vertebrates (fish, amphibians, birds, reptiles, and mammals) have declined

58 percent between 1970 and 2012, due largely to habitat loss and human-induced climate change. Freshwater populations were the hardest hit, dropping to less than 20 percent of their 1970 levels. Within the next three years (i.e., in 2020), it is thought that this planet will have lost two-thirds of the vertebrate biodiversity present in the mid-twentieth century (WWF 2016).

Science is warning that "we are at risk of being our own assassins" (Raff 2012). There is no safeguard for us if we cannot use our brains well. Science cannot fight toxic industry alone; it needs the moral, grassroots, soul-mobilizing ability that can only be found in the religious impulse. Science and religion are the estranged grandchildren of wonder. Science and religion need to form alliances to preserve the wonder of this world; they need to form alliances to preserve the creatures of this world; and they need to form alliances to keep alive the curiosity and the awe that allow their own renewal. And they need to do this now. One medieval Rabbinic commentary (Midrash Rabbah, a commentary on Ecclesiastes 7:13) has God showing Adam the glories of Creation, saying, "Take care not to spoil or destroy my world. If you ruin it, there's no one to repair it after you."

We are multiplying so fast that we are overspilling the livable acreage of the planet. Moreover, our waste products are destroying the other parts of the planet, as well. We demand space for housing, space for growing food, space for manufacturing, and space for disposing our waste. As humans take over more of the earth for themselves and these functions, they destroy the habitats of others. As Thomas Huxley (1894) noted, "Man shares with the rest of the living world the mighty instinct of reproduction and its consequence, the tendency to multiply with great rapidity." As the health of society improves, allowing more people to live to reproductive age, the more people will multiply. As mentioned, we are at the point at which the earth, as we have known it, is disappearing. In place of nature, in place of community, a "great plantation" is forming (Haraway 2015). Genes, seeds, livestock, and peoples are being forcibly moved from their place of origin to places where they can be regimented and controlled. The controllers get wealthy. The controlled at least get fed and entertained. Western religions, however, have kept telling people that it is God's will that we keep multiplying.

Can we now say that we've fulfilled that command and that other concerns must take precedence?

Birth control and family planning should allow us to decide how many children we have. Moreover, we should consider that families can be restructured such that each child has five or six parents, most of them nonbiological. The concern over genes and biological descendants has gone from being a source of pride to being an unbearable and dangerous neurosis. If children are to be precious, it will become critical to have fewer of them. Only that way can the earth recover from what we have already done to it. It also comes down to reproductive physiology and fertility, doesn't it? We call ourselves *Homo sapiens*, the "smart people." How we deal with climate change and overproliferation will show whether or not that name is justified.

THE SPELL OF THE GENE

The ad for Ancestry.com's DNA testing company is not only selling me a service, it is selling me an ideology: It tells me that if I give them some money and some of my DNA, I could "answer, once and for all, what it is that makes you you." Really? DNA makes me me? I'm sorry, but this is not science. It is a spell. DNA is the score, not the performance. Who we are—as opposed to what we are—depends on many things besides genes—our upbringing, education, what's in our environment, opportunities for physical and intellectual growth, and just plain dumb luck.

Twins start off with the same DNA, but they can become quite different (Fraga et al. 2005). I was at the Mütter Museum in Philadelphia when Stephen J. Gould filmed a television broadcast in front of the plaster-of-Paris death cast of Eng and Chang Bunker, the "original" conjoined ("Siamese") twins. Eng and Chang were very different people—Chang became an aggressive fellow who liked strong drink, whereas Eng was a mild-mannered teetotaler. And these two businessmen shared not only a liver and a circulatory system; they shared their genomes (they were identical twins), and, by force, they had to share the same environment. What makes us "us" is a very complex mixture of genes, environment, and experience.

But we hear over and over again, "DNA makes us who we are." As mentioned in Chapter 1, we are even told this by the car ads in our magazines, the genetic testing company ads on our televisions, and the anti-Choice websites on our computers: We are what our genes dictate. DNA had come to be seen as our essence, perhaps even our soul (Nelkin and Lindee 1996).

We are starting to believe our advertising. Former governor of Arkansas Michael Huckabee (2015) recently claimed, "We clearly know that that baby inside the mother's womb is a person at the moment of conception. The reason we know that it is is because of the DNA schedule that we now have clear scientific evidence on." However, there's no such "schedule." Similarly, another candidate for the 2016 Republican presidential nomination, Carly Fiorina (2015) stated, "Science is on our side. It shows . . . the DNA on the day that we die is the same DNA we had as a zygote." Actually, it isn't. There's no such thing as a "DNA schedule," and the DNA that we die with is different from the DNA with which we come into the world. Identical twins start off with the same DNA, but as they get older, their DNAs diverge (see Gilbert 2015a; Gilbert and Epel 2015). The DNA become modified by experience and by chance.

The science of epigenetics has told us this over the past twenty years. We can see the effects of environment on the DNA of laboratory animals. The DNA of genetically identical rats, for instance, is altered by whether or not the mother rat gives them attention during the first week of life. Certain genes become methylated (having small organic molecules attached to them), and the result is a marked change in anxious and sexual behaviors. In genetically identical mice, DNA is altered by chemicals (including food) that the mouse experiences while in the uterus. And this exposure, too, can have both physical and behavioral consequences (see Feil and Fraga 2012; Mitchell et al 2015, Gilbert and Epel 2015).

As discussed, DNA is altered by experience, and what we receive at fertilization does not predict who we will be. It is not our "essence." It is certainly not our soul. Fertilization is when we get our DNA. And while DNA may restrict our potentials in certain ways (I will never become six-foot-five, no matter how much I exercise), it does not tell us who we will become. This DNA provides the instructions to build our hearts and guts. It makes sure our eyes are only in our heads and not in our butts. And it constructs our brains, these remarkable organs that can learn and change, organs that allows us, as geneticist Barton Childs (2003) noted, "to escape the tyranny of our genes."

We are being told that our genome is the most important thing we can transmit. It is "who we are." Really? Did your genome give you a sense of responsibility? A sense of humor? A work ethic? A joie de vivre? A faith? An ability to love? Or did parental guidance, friends, family, a chance meeting with a remarkable person or place? Genes are critical in restricting what we can become, but they do not make us who we are.

CODA

So let's put this together: The wonder of life, the fear of extinction through overpopulation, and the spell of genetic transmission. Are the assisted reproductive technologies (ART) the ones we wish for? Or is it our society that makes us think so? We are given many ART options, but not many options to opt out of ART.

I was a minor casualty of the Icelandic volcano eruption in 2010. I had been working in Helsinki, and I could not return home. As long as there was no way possible of getting home (since no planes were flying), I could accept it. My colleagues and I had a volcano party, at which we had volcanic drinks and I taught the Finns Jimmy Buffett's volcano song. But as soon as flights became available, I had a responsibility to get home as soon as possible. On a much more important scale, if a couple knows that having a child naturally is not ever possible, it is something the couple could live with. One's body has failed, and there will be stigma; but what can one do about it? But as soon as there become possibilities for becoming fertile, then one feels obligated to try and try and try. Even if it bankrupts the couple. Even, as in the case of the couple Clara mentioned in chapter 2, it exiles them.

One has a great and expanding "choice" of ART procedures, but very little choice *not* to use them. This is part of what my friend and colleague Barry Schwartz (2004) has called "the tyranny of choice."

So to be better able to "choose" not to use these technologies, one must be able to recognize the spells, and one of these spells is that the genes constitute our soul and that we get this soul at fertilization. Another is to recognize the spell that tells us that the only way to have a family is heterosexual and genetic. This spell is beginning to unravel, and this permission to think in terms of other models of family may be the greatest contribution of the LGBT (lesbian, gay, bisexual, and trans) community.

Is the "nuclear family"—the male husband, female wife, and their biological children—the best we can do? Let's think of alternatives. Insisting that motherhood is *not* the be-all and end-all of a woman's life and that a woman's reproductive freedom is more important than societal dictates that demand reproduction, philosopher Donna Haraway (2016) has recently urged a total rethinking of what a family is. In light of the enormous and unsustainable increase in the human population (in 1970, there were roughly half as many people in the world as there are today), the idea that nuclear families are the ultimate units of child production, national identity, and cultural force, has to be called into question. Haraway (2016, 6) writes,

> Food, jobs, housing, education, the possibility of travel, community, peace, control of one's body and one's intimacies, health care, usable and woman-friendly contraception, the last word on whether or not a child will be born, joy: these and more are sexual and reproductive rights. Their absence around the world is stunning. For excellent reasons, the feminists I know have resisted the languages and policies of population control because they demonstrably often have the interests of biopolitical states more in view than the well-being of women and their people, old and young.

Is this view radical? Actually, it was recommended by the current Dalai Lama in 1993 (Dalai Lama 1996, 7). In a series of interviews, he advocated for pre-conception birth control: "But we are now confronted with an excess of precious lives, with far too large a world population. When it comes down to choosing between the survival of humankind as a whole and a few potential human births, the necessity for implementing birth control becomes obvious." Indeed, the Dalai Lama (1996, 79) sees overpopulation as one of the most important factors promoting violence.[1]

Population control *is* the issue. And to implement it without falling into racism, nationalism, imperialism, classism, and religious fanaticism is the duty of present generations. New ideas must be generated. But to replace the old with the new, the myths underlying some of the older and more harmful ideas must be exposed. Perhaps they will be replaced by other myths, but these will be myths for our times, not for our Bronze Age ancestors. Old and nonfunctional myths, especially those embedded in our science, will have to be exposed so that something better can be built, and myths about reproduction are among the hardest to relinquish, even when exposed.

We must alter our myths, as well as our tax incentives, to favor and celebrate low birth rates and the formation of large families with a small number of children: more adults and pets, fewer kids. In addition to the biological family, nongenetic-kin families would be encouraged. (This nongenetic component would most likely include other species, not only our microbial symbionts, but dogs and perhaps even horses, the two species that have had the closest familiar relations throughout history.) Adopting the elderly might become a common, even expected, part of one's family dynamics (Haraway 2016).

Birth should never be a matter of coercion. In either direction. No one should be forced not to have a child, but no one should be forced—by economic or social forces—to have children, either. As Natalie Angier (1999b) has pointed out, it is quite easy to love any child, and the supremacy of the gene has, in the past fifty years, made many people feel that it is even more important to have a child that is "built from the beginning" by its parents. This is a myth, a spell, that needs to be known, eliminated, and countered. Despite outmoded theories (such as those of Dawkins [1976]) claiming that one will sacrifice one's life only for close relatives (who share one's DNA), the scientific data say otherwise. The "selfish gene" is one of those spells. Humans are among the most altruistic of species, willing to help others who are not closely related at all. Moreover, we probably became that way through "cooperative breeding," the gathering of mothers together to help care for their infants (Balter 2014; Burkart et al. 2014).

However, the people of numerous countries, ethnicities, and religions are continually being told to reproduce their kind. People in many countries are given tax breaks for having children, and citizens of several European nations, as well as Japan, have been exhorted to have more babies to counter a slower birth rate and thus a lower percentage of the world's population. As I'm writing this, the Italian Ministry of Health is sponsoring "Fertility Day" festivities, telling citizens that "beauty has no age; but fertility does" and that a low birth rate is dangerous for the country's future (Mei 2016; Zilman 2016). This is the flipside of racism, a side that demands one's culture be continued though its genetic progeny.

But we are at an amazing point in our life history as a social species. Just as our biological propensity to reproduce is being inflamed by social and technological factors, our means of cultural reproduction has expanded far beyond its original genetic confines. Thanks to television and the Internet, Americans can learn the Maori haka-haka dance, Swedes can eat a kimchi recipe that had been passed down through a Korean family for seven generations, and Nigerian teenagers can purchase Mickey Mouse t-shirts. Similarly, the tenets of Shingon Buddhism and Orthodox Judaism are being transmitted to thousands of people without their having to be born in these traditions. Cultural appropriation is the new normal. (And it has always been part of the *American* tradition.) Amazingly, during the time when ART has enabled the production nuclear families, that type of family is no longer the sole transmitter of cultural inheritance.

New ideas will demand that economists, political scientists, religious leaders, scientists, housewives, and artists, all think in new ways. And think together. And sacrifice ideas that no longer work. I certainly don't have the answers or even a single answer. I can only ask that we not close our minds to some possibilities that, may at first, seem bizarre. New ideas for the family, ones where multiple parents, not just two, exist for each child, are no more bizarre than the idea that a single cell can become all the different types of cell in the body and that that body can think and love. They are no more bizarre than the redefinition of motherhood brought about by ART. Who will benefit from this replacement of the genetic family? Probably our children and the children of our friends. Because they will gain a larger and more varied circle of adults to care for them. Adults will also benefit from increasingly flexible opportunities to work with and learn from others, and will be able to grow in relationships with children who may or may not be genetically related. Cooperative parenting has always been part of the human experience, and (as all teachers know), loving other people's children not only defines humanity, it is about the easiest thing in the world.

A FIELD GUIDE TO ASSISTED REPRODUCTIVE TECHNOLOGIES

SCOTT GILBERT

Happy is that person who can discern the causes of things.

—Virgil, *Georgics*

We've heard about some of the assisted reproductive technologies (ART) in general and how they can affect people's lives. This appendix is for those interested in going a step deeper into the different ART techniques. Here is *The Talk* for young adults in the 21st Century. While not for the expert or the physician, this appendix will introduce some of the major steps of in vitro fertilization and its variations, including intracytoplasmic sperm injection, gamete intrafallopian transfer, zygote intrafallopian transfer, and mitochondrial replacement therapy. It will briefly explore some of their success and shortcomings. It will also discuss preimplantation genetic diagnosis as an emerging technology that may enable people to have a child of their desired sex and free of certain genetic diseases.

As we've seen in the preceding chapters, infertility can be caused by several problems. It might occur by the failure to ovulate a mature oocyte, by having too few functional sperm, by a physical blockage of the male or female ducts, or by incompatibilities between the sperm and the milieu of the egg or the reproductive tract. While there are numerous treatments for women that can lead to the maturation and ovulation of oocytes, there are relatively few treatments for men who are not making sufficient sperm. In women, drugs that increase estrogen levels, such as exogenous gonadotropins, or antiestrogenic drugs, such as clomiphene and tamoxifen, can be used to stimulate the ovaries. In men, sperm may be concentrated and injected either into the oocyte or the reproductive tract near the oocyte.

There are numerous types of ART to circumvent infertility. The simplest is **artificial insemination** (**AI**). This is the mechanical introduction of sperm into the female reproductive tract for the purpose of achieving pregnancy. Artificial insemination is widely used for relieving infertility when the man is not fertile (when the man produces too few sperm, too few motile sperm, or is unable to have an erection) or when a woman has a blockage in her reproductive tract and sperm have to be placed on the other side of the blockage. Artificial insemination has also been useful in cases of unexplained infertility and to allow lesbian couples and single women to have children. Basically, sperm from a donor male is injected (through a tubular instrument called a catheter) into the woman's reproductive tract after the woman has been detected to be ovulating (by temperature, hormone levels, or ultrasound).

The sperm can be from the husband, but it can also be from sperm donors, either arranged or purchased. Many religions prohibit AI because they believe that it weakens the bonds uniting the act of sex and the process of procreation. Some religions also proscribe donor AI because the child would be biologically that of the donor, even if raised by the social parent.

In vitro fertilization (**IVF**) is an infertility treatment in which eggs and sperm are retrieved from the female and male partners and placed together in a Petri dish for fertilization. After the eggs have begun dividing, the embryos are transferred into the female partner's uterus, where implantation and embryonic development occur as in a normal pregnancy.

As has been discussed, IVF was developed in the early 1970s to treat infertility caused by blocked or damaged fallopian tubes. The first IVF baby, Louise Brown, was born in England in 1978. Since then, the number of IVF procedures performed each year has increased, and success rates has improved significantly (box AP1.1). Success rates compare favorably to natural pregnancy rates in any given month when the woman is under age forty and there are no sperm problems (Trounson and Gardner 2000; Bavister 2002).[1]

BOX AP1.1: SUCCESS RATES AND COMPLICATIONS
OF IN VITRO FERTILIZATION

It has been mentioned more than once in these pages that the success rate of IVF and other derived procedures are variable, unclear, easily contradictory, and often dependent on the source of the information. This has caused a huge amount of frustration, especially for older women whose fertility is much lower than women in their twenties (see Zoll 2013). It is thus appropriate to include here some further data on these matters. The rate of delivery of live babies per oocyte retrieval depends on the age of the female partner. Of the 173,200 cycles of ART conducted in 2014, some 57,323 produced a live birth. That's a success rate of about 33 percent. This rate compares well with the one-in-four (25 percent) probability of achieving conception during each cycle for normal healthy couples having unprotected intercourse. Thus, IVF offers improved chances of conception to some infertile couples. The success rate for the percentage of cycles giving rise to a live baby drops to 25.5 percent, however, for women thirty-five to thirty-seven years of age, and to 17.1 percent for women aged thirty-eight to forty. After forty years of age, the success rate is less than 5 percent (Lipshultz and Adamson 1999; Speroff and Fritz 2005; CDC 2016). This decline is most likely a result of the higher frequency of chromosomal and biochemical abnormalities of eggs as women advance in age. About 1.6 percent of babies in the United States are born through ART.

Although the IVF procedure is quite successful in achieving pregnancy, it does carry the risk of multiple births. In 2012, 36 percent of ART outcomes using fresh eggs and sperm produced pregnancies. Of these, about 30 percent were multiple-fetus pregnancies (CDC 2012). The rate of multiple births depends on the age of the woman and the number of embryos transferred. When three embryos were transferred, the multiple-birth rate was 46 percent for women aged twenty to twenty-nine. The rate was only 39 percent for women aged forty to forty-four when seven or more embryos were transferred. Multiple birth is a serious concern because multiple-birth infants are predisposed to many health problems, including premature delivery, congenital mal-formations, infant death, and low birth weight (Lipshultz and Adamson 1999; Schieve et al. 1999; Bhattacharya and Templeton 2000). Babies born prematurely and at low birth weight are at risk for cerebral palsy and chronic respiratory problems. In addition, mothers who carry mul-tiple infants are also at risk for many health conditions and complica-tions (e.g., high blood pressure, diabetes), and the costs associated with multiple pregnancies are also greatly increased. The costs of delivery and hospitalization for the first five years is significantly higher for multiple-birth babies (van Heesch et al. 2015).

Because IVF was the first assisted reproductive technology to be widely publicized, many people mistakenly believe that IVF is the only treatment option for infertile couples. Actually, most infertile couples respond well to less complicated treatment options, such as hormonal therapies and AI. However, IVF remains one of the most commonly used of the ART procedures.

THE IVF PROCEDURE

The IVF procedure has four basic steps (NLM 2016) (figure AP1.1):

1. *Ovarian stimulation and monitoring.* Having several mature eggs available for IVF increases the possibility that at least one will result in a pregnancy. Typically, gonadotropins or antiestrogens are used to "hyperstimulate" the ovaries to produce several mature oocytes. As mentioned, this is not a simple procedure. The hormones used to stimu-late the production of numerous oocytes cause nausea and vomiting in about a third of the women having this procedure. However, in some women, persistent vomiting, severe abdominal pain, rapid weight gain, and even life-threatening respiratory distress can occur. The risk of such problems is greater in younger women experiencing multiple rounds of hormone therapy (ASRM 2008).

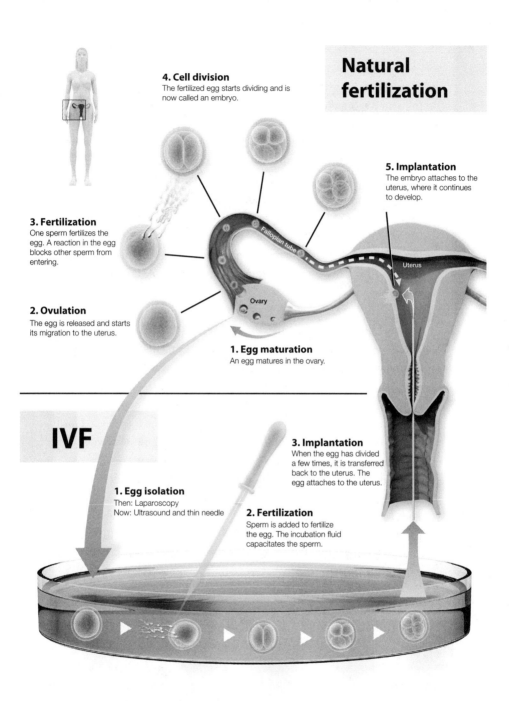

4. Cell division
The fertilized egg starts dividing and is now called an embryo.

Natural fertilization

5. Implantation
The embryo attaches to the uterus, where it continues to develop.

3. Fertilization
One sperm fertilizes the egg. A reaction in the egg blocks other sperm from entering.

2. Ovulation
The egg is released and starts its migration to the uterus.

1. Egg maturation
An egg matures in the ovary.

Fallopian tube

Uterus

Ovary

IVF

1. Egg isolation
Then: Laparoscopy
Now: Ultrasound and thin needle

2. Fertilization
Sperm is added to fertilize the egg. The incubation fluid capacitates the sperm.

3. Implantation
When the egg has divided a few times, it is transferred back to the uterus. The egg attaches to the uterus.

FIGURE AP1.1 Comparison of Natural Fertilization with In Vitro Fertilization.

Source: Illustrated by Mattias Karlén. Courtesy of the Nobel Commission.
Copyright © Nobel Committee for Physiology or Medicine.

2. *Egg retrieval.* Once the follicle has matured (but has not yet been released from the ovary), the physician attempts to retrieve as many eggs as possible. The female gametes about to be ovulated are actually oocytes that have reinitiated cell division due to the hormones. The physician retrieves the oocytes by guiding an aspiration pipette to each mature follicle in the ovary and sucking up the oocyte. "Harvesting" eight eggs means making eight small punctures in the ovary. Once the oocytes are recovered, those that are mature and healthy are transferred to a sterile container to await fertilization in the laboratory. Initially, this was done through the abdomen, with the patient placed under general anesthesia. But it is now most commonly done by inserting a needle through the upper vagina.

3. *Fertilization.* A semen sample is collected from the male partner approximately two hours before the female partner's oocytes are retrieved. These sperm are then processed (a procedure called sperm washing). Sperm washing occurs in a medium that artificially capacitates the sperm. The healthiest looking and most active sperm in the sample are then placed into the Petri dish with the oocytes, and the gametes are incubated at body temperature. In general, each oocyte is incubated for twelve to eighteen hours with fifty thousand to one hundred thousand motile sperm. The success rate for fertilization is between 50 and 70 percent. If fertilization is successful, the eggs will begin to divide, and the resulting embryos will shortly be ready to be transferred into the uterus.

4. *Embryo transfer.* Embryo transfer is not a complicated procedure and can be performed without anesthesia. It is usually done three days after egg retrieval and fertilization. The physician looks for healthy embryos (those that have divided well, containing six to eight cells). These embryos are sucked into a catheter. The physician then inserts the catheter through the female partner's vagina and cervix to transfer the embryos directly into the uterus. Normal implantation and maturation of at least one embryo is required to achieve pregnancy.

In cases in which fertilization has been achieved in vitro, but after a number of cycles, implantation into the uterus fails, the physician may suggest "assisted hatching," in which a small hole is lysed in the zona pellucida prior to inserting the embryo into the uterus. The zona pellucida is

that protein coat surrounding the early egg and embryo, and this procedure ensures that the embryo will be able to hatch from the zona pellucida
in time to adhere to the uterus.

VARIATIONS ON IN VITRO FERTILIZATION

In addition to the sperm-meets-egg-in-a-dish way of performing IVF,
other techniques have become available. **Intracytoplasmic sperm injection (ICSI)** was developed to treat couples who had a low probability of
achieving fertilization due to the male partner's extremely low numbers
of normal viable sperm. Male infertility may be caused by genetic factors
leading to poor sperm production or by blockages or abnormalities in the
ejaculatory ducts. Men who have had a severe injury to their reproductive
organs, a vasectomy, or chemotherapy or radiation for testicular cancers
may have few sperm in the ejaculate. In ICSI, a single sperm is injected
into the cytoplasm of an egg. The sperm cell membrane unwraps, allowing
the haploid sperm nucleus to meet with the haploid egg nucleus. The thin
sperm cytoplasm can then activate the egg to begin development. When
combined with IVF, ICSI allows couples for whom low sperm count is an
issue a more favorable probability of achieving conception.

Preliminary data, however, suggest that in instances where ICSI is performed because of intrinsically low sperm counts (not because of trauma
or therapy), this procedure may just be moving the problem down to the
next generation. Belva and colleagues (2016) have shown that the sperm
count of a cohort of fifty-four young adult men conceived though ICSI
was about half that of naturally conceived men of the same age. Men conceived through ICSI were more than three times more likely than other
men to have sperm counts below 15 million sperm/ml (the concentration
below which men are considered to have low sperm counts).

Gamete intrafallopian transfer (GIFT) was developed in 1984 as
another variation on IVF. Here, sperm are injected into the oviduct at
the time when the oocytes are ovulated. This technique is often used in
couples with unexplained infertility in which the female partner has at
least one open fallopian tube. Gamete intrafallopian transfer has also
been recommended for couples whose infertility is due to cervical or

immunological factors that prevent the sperm from reaching the oocyte in the oviduct. The main difference between IVF and GIFT is that GIFT fertilization occurs naturally within the female partner's body, instead of in the laboratory as with IVF.

Zygote intrafallopian transfer (**ZIFT**) is another variation on IVF. This procedure is also called *tubal embryo transfer*. As in IVF, fertilization takes place outside the body in a Petri dish. The resulting zygotes are then transferred into one of the female partner's fallopian tubes. The location where fertilization takes place and the physician's ability to observe and confirm fertilization are the main differences between ZIFT and GIFT. With GIFT, the actual fertilization cannot be observed, because the eggs and sperm are united inside the female partner's fallopian tube.

Mitochondrial replacement therapy (**MRT**) is yet another variation on IVF. There are some genetic diseases that involve mutations in a cell's mitochondria. These are the "organs" of the cell that provide it with energy, and the sperm uses its mitochondria to reach the egg. When a sperm enters an egg, its mitochondria are typically used up, and if not, the sperm mitochondria are destroyed. It's only the egg's mitochondria that survive. So all our mitochondria come from our mom. Mitochondria have their own DNA, and there are thirty-seven genes in human mitochondria. About one person in ten thousand suffers from a mutation in one of these genes. Mutations in the mitochondrial genes can cause severe and progressive diseases of the eye, kidney, and musculature. These diseases are inherited solely from the mother, since only the egg provides them.

So an ingenious variation of IVF can prevent their transmission (Falk et al. 2016). This procedure involves the use of "three parents." First, the couple who wants a child undergo IVF, fertilizing the mother's egg (which has the mitochondrial mutation) with the father's sperm. Then, the sperm is also used to fertilize the egg of a "mitochondrial donor." As the sperm and egg nuclei approach each other in the mitochondrial donor, they are removed by a thin pipette that sucks them out of the egg. These nuclei are replaced by nuclei from the egg fertilized by the couple that wants a baby. In this way, the couple's nuclei are placed into an egg that has been activated to develop.

Alternatively, the nucleus of the mitochondrial donor's egg (containing half the number of chromosomes) can be removed prior to fertilization and replaced by the egg nucleus of the woman with the mitochondrial disease. This egg can then be fertilized. In either case, it contains a father's sperm nucleus, a mother's egg nucleus, and cytoplasm containing the healthy mitochondria of a second egg.

This remains a controversial procedure for several reasons. First, unlike most IVF procedures, the two methods used for MRT reconstruct the fertilized egg in radical ways, and they constitute a type of gene transfer. One can view MRT as replacement of mitochondria; or one can view it as the replacement of half a genome. Embryologist and bioethicist Stuart Newman (2013) claims that this is a slippery slope to make cloning and germline gene engineering more acceptable to the general public. Second, there are safety issues specific to this technique, since residual mutant mitochondria can possibly destroy the donor mitochondria. The first healthy child born from such procedures was conceived by circumventing the laws of the United States and was conceived in Mexico. The failure of international law on this issue has been of great concern as it may promote reproductive tourism. Those scientists who are against using this therapy claim that they do not help existing people, and if women did not want to transmit mutant mitochondria, other techniques, ranging from egg donors, preimplantation genetic diagnosis, and adoption are available (CGS 2016b).

PRENATAL DIAGNOSIS AND PREIMPLANTATION GENETICS

Last, there is the question of preimplantation genetics and the selection of the "right" embryo. One of the consequences of IVF and the ability to detect genetic mutations early in development is that a new area of medicine has arisen called preimplantation genetics. Preimplantation genetics seeks to test for genetic disease *before* the embryo enters the uterus. After that, many genetic diseases can still be diagnosed before a baby is born.

By using IVF, one can consider transferring only those embryos that are most likely to be healthy rather than aborting those fetuses that are most likely to produce malformed or nonviable children. This can be achieved by screening embryonic cells before the embryo is transferred in the womb. (Yes, it probably should be called "pretransfer genetics,"

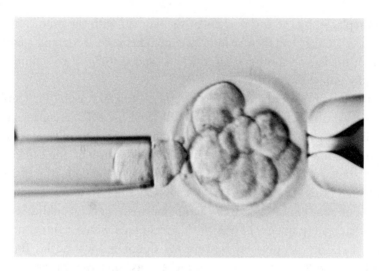

FIGURE AP1.2 Preimplantation Genetic Testing.

Preimplantation genetic testing begins with in vitro fertilization. Then, as shown here, a cell is taken from the embryo. The genes of this cell are then analyzed, while the embryo regulates and restores the missing cell.

but that sounds even more unnatural.) While the embryos are still in the Petri dish (at the six-to-eight-cell stage), a small hole is made in the zona pellucida, and two cells are removed from the embryo (figure AP1.2). Since the mammalian egg undergoes regulative development, the removal of these cells does not endanger the embryo, and the genes from these isolated cells can be tested immediately. Numerous molecular biological techniques can be used to determine the presence or absence of certain genes in the cell, and whether the normal number and types of chromosome are present. Results are often available within two days. Presumptive "normal" embryos can then be transferred into the uterus, whereas any embryos with deleterious mutations can be discarded.

SEX SELECTION AND SPERM SELECTION

However, the same procedures that allow preimplantation genetics also enable the physician to know the sex of the embryo. Knowing the sex of a six-to-eight-cell embryo conceived by IVF increases the possibility that parents could decide to have only embryos of a desired sex transferred into the uterus. Sex selection using preimplantation genetics is seen by many as a beneficial way of preventing X-linked diseases such as hemophilia, but, in fact, it is often used as a method of simply choosing your offspring's sex (box AP1.2). Opponents of sex selection point to its possible use to prevent the birth of girls in cultures where women are not as highly valued as men (Zhu et al. 2009). Different countries and even different hospitals have different policies permitting preimplantation genetic diagnosis solely for the purpose of sex determination. But in the United States, if one has the desire and the money, one can usually determine the sex of one's offspring before implantation.

BOX AP1.2: TECHNOLOGY AND THE PRESSURE FOR SEX SELECTION

Technologies can be major agents of cultural change, and this is very evident in the role of ultrasound. Many factors play into a parent's desire for children of one sex over the other, and these factors vary in different parts of the world. In most of the world—and particularly the East Asia—cultural, personal, and economic issues merge to drive the balance in favor of male babies (Robertson 2001). Much of this attitude is deeply ingrained in the cultures that share it, and many regions have long histories of preferring males. On the other hand, many Westerners who promote sex selection see it as a way to achieve what is known as family balancing, since, in these countries, it appears the preference in most families is to have children of both sexes (Kalb 2004).

Matters of finance and family economics are perhaps the most overt forces working around the globe to drive sex selection. In India, for example, daughters have traditionally been viewed as an expense, whereas sons are seen as a financial asset (Ramachandran 1999). In some parts of India, particularly poor rural areas, it is not uncommon for the midwife (*dai*) to hold a female newborn upside down by the waist, give it a jerk to snap the spinal column, and pronounce a stillbirth (Carmichael 2004). Among the Chinese, the kinship system emphasizes paternal descent.

Living in the father's home is the norm for couples, and Chinese parents rely on their sons for support in their old age. Because the perceived need for a son is so high, and because the Chinese government places heavy financial burdens on families with more than one child (a policy that ended in 2015), infanticide of females is believed to be relatively common. Female infants are often left in the streets or on doorsteps of orphanages (Li 1991; Vines 1993; Winkvist and Akhtar 2000).

The overwhelming preference for male children among parents in India and China has had sobering results. The ability to see the genitalia of a child through ultrasound now allows sex identification within the first trimester of pregnancy—well within the legal limits of abortion in most countries. If a woman requests an abortion, it is impossible in most cases to differentiate whether she does not want a baby at all or does not want a baby of a particular sex.

In wealthy areas of India, where ultrasound services are available, abortion rates of female fetuses are disproportionately high. In Bombay, a 1985 survey found that 96 percent of aborted female fetuses were aborted after amniocentesis revealed their sex (Ramachandran 1999). One study found that out of 8,000 reported abortions, 7,999 of them were of female fetuses (Roberts 2002). In 1994, the Indian government passed the Pre-Natal Diagnostic Techniques Act, in an attempt to regulate prenatal testing such as sonography and make it illegal to use such procedures for sex selection. However, it seems that sex-selective abortion following sonography remains widespread (Shete 2005).

The overwhelming preference for boys in most Eastern cultures, coupled with the inexpensive sex determination technology, has generated the gender gap—the extent to which the sex ratio of males to females deviates from the theoretical norm of 100:100 in many countries (Macklin 1995; Satpathy and Mishra 2000). It is possible that 160 million women (equal to the entire female population of the United States) are presently "missing" from the Chinese population (Hvistendahl 2011).

In some regions (most often rural areas), gender gaps have led to a generation of young, single males who have very few prospects of marrying. One report (den Boer and Hudson 2002) claims that 94 percent of all unmarried people aged twenty-eight to forty-nine in China are male, and that 97 percent of these men have not completed high school. This demographic group has been around long enough in some countries to earn its own term. In China, the young men are known as *guang gun-er* ("bare branch") because they represent "branches of a family tree that would never bear fruit because no marriage partner might be found for them" (Hudson and den Boer 2004). Studies have shown that the *guang gun-er* commit a disproportionately high fraction of the crime in their respective areas. Sex selection can have profound consequences.

BASIC QUESTIONS RAISED BY ASSISTED REPRODUCTIVE TECHNOLOGIES

Assisted reproductive technologies began as a way of allowing infertile couples to achieve pregnancy. These techniques have been successful in that by now their rates of delivery are considered just about equal to those achieved by normal fertilization. However, these technologies have raised several ethical and legal concerns (Purdy 2001; Gilbert et al. 2005). Here are some of the most recurrent and thus the most often mentioned concerns in this book, in an attempt to summarize the main problems we are facing right now when it comes to designing guidelines for the future:

- Whom do these technologies assist? Couples may pay $12,000 to $15,000 for a cycle of IVF in the United States. Couples often pay between $50,000 and $200,000 to achieve a single pregnancy (Andrews 1999; Caplan 2005; Uffalussy 2014). Is ART only for the wealthy? If a woman with infertility knows that she could possibly have a genetically related child if she were wealthy, does this frustrate more women than it helps? What can be done about infertility in poor countries with few resources? If curing infertility is our goal, then should our focus be on high-tech medicine or on public health efforts to eliminate sexually transmitted diseases (one of the leading causes of infertility)?
- Is there a "right" to have a genetically related child? Are procedures that can allow fifty-year-old women to have babies the best use of our

medical resources? Is the use of surrogacy carriers "right" for any and all reasons, if one can afford to pay for their services?

- Why is there a "need" to have genetically related children? Is this desire "biological," or is it being manufactured by the advertising done by fertility clinics competing with one another in the present market?
- What is the status of a frozen embryo? Is throwing away the extra embryos produced by IVF equivalent to abortion? Who has the right to keep the embryos if a couple should divorce? (And is the biological father obligated to make child support payments if the embryo is transferred and comes to term?) It is estimated that there are hundreds of thousands of frozen embryos currently in storage.
- Are current ART procedures safe for mothers and offspring? While the link between hormones and reproductive cancers has been known for years, it is not known whether women undergoing extensive cycles of ART are at risk for cancers. Moreover, IVF procedures grow early embryos in artificial media, and animal studies have shown that diet during this preimplantation period (before pregnancy) is important for the normal methylation of particular genes and for normal development and postnatal health. The consequences of such artificial nutrition on adult human health and behavior are only starting to be addressed.
- How should infertility clinics be regulated? In contrast to the United Kingdom, where there are strict laws regulating what infertility clinics can do and how they must report their results, infertility clinics in the United States are not under federal or state regulations. It is often difficult to compare success rates and health records among clinics. Indeed, what is an "experimental" treatment in one hospital might be considered "standard" treatment in another.

There are, as one author has stated (Erb 1999), more regulations on tattoo parlors than on fertility clinics. Assisted reproductive technologies are powerful technologies. As such, they have been a godsend for numerous couples, allowing them to experience the birth of a child they could never have had naturally. As such, they have also been a path to profound despair and financial ruin for couples for whom this remarkable scientific promise went unfulfilled.

GLOSSARY

ABORTION. The termination of pregnancy before the fetus can survive outside the uterus. Spontaneous abortions are called miscarriages (q.v.). Abortions done deliberately are called induced abortions.

ABORTUS. A dead or nonviable fetus; the product of an abortion (q.v.).

AI. See ARTIFICIAL INSEMINATION.

AMNION. The "water sac" that protects the embryo. It is derived from the embryo and contains fluid that acts as a shock absorber and a means of avoiding drying out.

ANALOGY. A comparison of the similar features of two things or between two elements in a relationship.

ART. See ASSISTED REPRODUCTIVE TECHNOLOGIES.

ARTIFICIAL INSEMINATION (AI). The injection of semen into the female reproductive tract by means other than copulation.

ASSISTED REPRODUCTIVE TECHNOLOGIES (ART). Fertility treatments used to achieve pregnancy. These include artificial insemination, in vitro fertilization, embryo freezing, among others.

BABY M. The pseudonym given to the infant born to a surrogate mother who successfully sued for the right to retain the infant.

BISPHENOL A (BPA). An organic compound that has been found to block meiosis in eggs and has been linked to human infertility. It is very common in plastics.

BLASTOCYST. A fluid-filled early-stage mammalian embryo, consisting of an outer ring of cells that will become the chorion portion of the placenta (q.v.) and, attached to that ring, an inner cell mass (q.v.), which will become the embryo.

BPA. See BISPHENOL A.

CAPACITATION. A series of physiological events that change the membrane of the sperm such that it gains the ability to find and fuse with the egg.

CEREBRAL PALSY. A congenital (q.v.) syndrome of neurological disorders, including limitation of movement and postures and increased muscle tone. Intellectual development is not compromised. The symptoms vary from one person to the next, and, while the causes are not known, premature birth is a large risk factor.

CERVIX. The lowest portion of the uterus, connecting it to the vagina. The cervical mucus helps regulate sperm flow to the uterus.

CHORION. The fetal portion of the placenta (q.v.). It is derived from the outer cells of the blastocyst and adheres to the endometrial cells (q.v.) of the uterus.

CHROMOSOME. An organized array of DNA (q.v.) and proteins. Humans usually have forty-six chromosomes in each cell. They are the physical material of genes.

CLEAVAGE. The stage of early development following fertilization (q.v.) characterized by rapid cell division.

CLONED ANIMAL. An animal produced by the transfer of the nucleus from one cell into the activated enucleated oocyte (q.v.) of another animal, such that the newly implanted nucleus directs normal development.

CLUSTERED REGULARLY INTERSPACED SHORT PALINDROMIC REPEATS. See CRISPR.

CONCEPTUS. The product of fertilization at any stage of development: zygote, embryo, fetus, newborn, and adult.

CONGENITAL. Describes a condition existing at or before birth.

CONGENITAL ANOMALIES. Birth defects. Departures from normal development seen when the baby is born.

CONTRACEPTIVE PILLS. Pills containing chemicals, often hormones, that prevent the sperm and the egg from meeting. They do not cause abortions.

CORONA RADIATA. The cells surrounding the egg when it is ovulated (q.v.). These cells used to be part of the ovary.

CRISPR (CLUSTERED REGULARLY INTERSPACED SHORT PALINDROMIC REPEATS). A stretch of DNA that evolved as part of the immune system of bacteria. It can recognize and eliminate particular sequences of DNA. The naturally occurring CRISPR has been modified for use as a tool to change DNA sequences to ones that are desired.

CYCLE. A menstrual cycle. In ART (q.v.), a "stimulated cycle" is one in which the woman has been given drugs to "hyperstimulate" her ovaries to produce more eggs.

DEOXYRIBONUCLEIC ACID (DNA). The chemical component of our genes, and thus, the physical basis of our heredity. The sequence of chemicals in our DNA instructs the composition of our proteins and where in which cells they are made.

"DESIGNER BABY." A slang term for a baby whose genes have been modified toward a desired end. Although babies have been selected for their sex or for their having certain genes, or another person's healthy mitochondria (q.v.), transgenic (q.v.) babies, whose genes have been altered and then selected, have not yet been born.

DIFFERENTIATION. The processes by which an unspecialized cell, such as the zygote or the cells of the inner cell mass (q.v.), becomes the more specialized cells of the body.

DNA. See DEOXYRIBONUCLEIC ACID.

DOLLY. A sheep, born in 1996, Dolly was the first mammal to be cloned using the nucleus of an *adult* cell for nuclear transfer.

ECOLOGICAL DEVELOPMENTAL BIOLOGY. The science that studies the relationships between developing organisms and their environments.

ECTODERM. The outermost of the three layers that generate the embryo, forming the skin, brain, and nervous system.

EEG. See ELECTROENCEPHALOGRAM.

EGG. The mature female gamete. Sometimes called by its Latin name "ovum."

EGG DONATION. The process by which a woman provides eggs for another person's ART or for medical research. Often, the woman donating the eggs is given hormones so that more than one egg can be matured and retrieved.

EGG FREEZING. Oocyte cryopreservation. Usually similar to egg donation (q.v.), except that a woman donates the eggs to herself, freezing them until she wishes to use them for in vitro fertilization (q.v.).

EGG RETRIEVAL. Part of the procedure for in vitro fertilization. After the woman is given drugs to stimulate the maturation of numerous eggs in her ovaries, the woman is put under anesthesia, and a needle, guided by ultrasound (q.v.), is inserted into the woman onto the ovary, where it sucks up the eggs that have matured but have not yet been ovulated.

ELECTROENCEPHALOGRAM (EEG). A record of the electrical activity of the brain. Measuring brain wave patterns, the EEG integrates electrical activity from all brain regions. Its loss is colloquially called "flatlining" and is characteristic of death.

EMBRYO. A developing organism before it is born or hatched. In humans, it refers to the first eight weeks of development, when the organ systems are beginning to form. (Compare with "fetus.")

EMBRYONIC STEM CELL. Cells derived from the inner cell mass of the blastocyst (q.v.) that have the potential to become any cell type in the adult body. This characteristic is called pluripotency (q.v.). As stem cells, they divide to form both another embryonic stem cell as well as a more differentiated cell type.

EMBRYO TRANSFER. Part of ART procedures in which the embryos that developed from in vitro fertilization have divided two or three times and are then placed into the uterus of the woman in the hopes they will implant and produce children.

ENDODERM. The innermost of the three layers that generate the embryo, forming the stomach, intestines, pancreas, liver, and lungs.

ENDOMETRIUM. The epithelial lining of the uterus. During the early stages of the menstrual cycle, it makes new cells that can adhere to the blastocyst (q.v.) and makes new blood vessels. If no embryo adheres, this is the tissue that gets sloughed off at the end of each menstrual cycle.

ENSOULMENT. In religious writings, this is the process in which a supernatural soul enters or is formed within the body.

ENUCLEATED. The state of a cell once its nucleus, which houses the vast majority of its chromosomal genes, has been removed. In cloning procedures, an enucleated oocyte is an oocyte that can accept another cell's nucleus and, when activated (via an electric shock in place of a sperm), can begin development.

EPIGENETIC REPROGRAMMING. The erasure and addition of methyl groups (q.v.) and other regulatory molecules during mammalian development. This happens naturally, as methyl groups are added and deleted as the undifferentiated cells of the embryo become specialized. This can also be done during cloning, when the methylation of the nuclear DNA must be largely erased so that the nucleus will act as if it were from a more pluripotent cell.

EPIGENETICS. The study of how cells acquire their specialized identities (e.g., neurons, gut cells) without changing their DNA sequence. These mechanisms usually involve the regulation of gene expression, determining which genes will be active in a particular cell.

ESTROGEN. A set of steroid hormones needed for the completion of female body organization and appearance. Estrogen is also needed for preparing the uterus for pregnancy. In both sexes, estrogens are used to coordinate bone growth.

EUGENICS. The social program promoting the betterment of the human species by selective breeding of those people with desired traits and the reduced breeding (through sterilization or economic incentives) of those with less desirable traits.

FALLOPIAN TUBES. See OVIDUCTS.

FERTILIZATION. The fusion of sperm and egg allowing the activation of development, followed by the fusion of their respective nuclei. This creates the zygote (q.v.).

FETUS. The stage of mammalian development between the embryonic stage (the first eight weeks in humans) and birth. It is characterized by growth.

FIBROIDS. Noncancerous growths that can arise from the uterus. While they rarely become malignant, they can cause infertility by blocking the sperm's path through the uterus or by blocking the early embryo's ability to implant (q.v.).

FOLLICLE-STIMULATING HORMONE (FSH). A protein hormone secreted by the pituitary gland that promotes the development of an oocyte in the ovary at the beginning of each cycle. It is also needed for sperm development in the testes.

FRATERNAL TWIN. A dizygotic (two-egg) twin. Twins that result when two eggs are matured during a monthly cycle in the woman. Two eggs are ovulated, and each comes into contact with a different sperm. Thus, two embryos are generated during the same cycle.

FSH. See FOLLICLE-STIMULATING HORMONE.

GAMETE. A mature sex cell, the sperm or the egg.

GAMETE INTRAFALLOPIAN TRANSFER (GIFT). A technique of ART (q.v.) for circumventing fertility, in which eggs are removed from a woman's ovaries, placed in one of her oviducts (a Fallopian tube), and mixed with a man's sperm.

GASTRULATION. The series of movements in the early embryo, whereby cells acquire new neighbors. Pluripotency is lost, as the new cellular interactions limit the fates of each group of cells. Sometimes called "individuation," as this is the point at which the ability to form identical twins ends.

GENE. A physical and functional unit of inheritance. Genes are made of DNA, whose sequence of nucleotides (four small specific molecules) determines the sequence of amino acids in a protein. The amino acid sequence in the protein determines the protein's functions. Each genes is located at specific a site on the chromosome.

GENETIC ENGINEERING. The deliberate alteration of the characteristics of an organism by manipulating its genome. (See also GENOME EDITING and GERMLINE GENE THERAPY.)

GENOME. The complete set of genetic material present in a cell. Usually, this includes the DNA of the nuclear chromosomes (about twenty thousand genes) and the mitochondria (about thirty-seven genes) in the cell cytoplasm.

GENOME EDITING. Sometimes called "gene editing," this is a type of genetic engineering in which DNA can be altered, added to, or removed from a person's genome (q.v.).

GERM CELLS. The lineage of cells that produces the gametes (the sperm and the egg). In mammals and insects, these make up the group of cells set aside for producing the next generation. ("Germ" comes from "germinal," that which is earliest).

GERMLINE GENE THERAPY. The treatment of disease by transferring "normal" DNA into the gamete precursors of people carrying a genetic disease. This would enable the correction of disease-causing genetic variants. Somatic cell gene therapy would replace deficient genes with new ones within an existing body (but these genes would not be in the sperm and eggs, and therefore would not be passed on to the next generation).

GIFT. See GAMETE INTRAFALLOPIAN TRANSFER.

GONADS. The ovaries and testes.

HAKA HAKA. A traditional war dance of the Maori people of New Zealand. Now performed by some New Zealand athletic teams during international competitions.

hCG. See HUMAN CHORIONIC GONADOTROPIN.

HUMAN CHORIONIC GONADOTROPIN (HCG). A hormone produced by the embryo after implanting into the uterus. It is then made by the embryonic portion of the placenta and instructs the ovaries to make progesterone, the hormone that enables the continuation of pregnancy. Often used in tests to confirm that a woman is pregnant.

HUNTINGTON'S DISEASE. A fatal genetic disorder that causes the progressive deterioration of the nerves in the brain. It results in the impairment of thinking and movement and ultimately causes death. It is caused by mutations in an autosomal-dominant gene, but as it often manifests only after age thirty, afflicted individuals may not know that they are affected until after having children of their own.

ICSI. See INTRACYTOPLASMIC SPERM INJECTION.

IDENTICAL TWIN. A monozygotic (one-egg) twin. Twins formed by a single fertilization event, wherein the early embryo separates into two distinct groups of cells. Therefore, each twin has the same genetic identity.

IMPLANTATION. The binding and penetration of the uterine tissues by the early embryo. The human embryo hatches from the zona pellucida on about the fifth day after fertilization and adheres to the uterine endometrium. The outer portion of the embryo (the trophoblast) then expands and digests its way into the uterus.

IN VITRO FERTILIZATION (IVF). A series of procedures designed to mix sperm and egg in a test tube so that they readily fuse and begin embryonic development. Eggs must be matured in relatively large quantities (through hormones), removed from the ovaries (by suction), and mixed with sperm.

INBREEDING. The production of offspring from the mating of closely related individuals. Inbreeding heightens the chances that rare genes will become expressed because they were received from both parents. It is not uncommon where people, animals, and plants live in regions isolated from the rest of the population.

INDUCED PLURIPOTENT STEM CELL (IPSC). Adult cells that have been converted to cells with the same developmental potency as embryonic stem cells.

Thus, the adult cells become pluripotent. This can be done by adding viral DNA-containing genes that activate the expression of those genes usually seen active in the inner cell mass.

INFERTILITY. The inability to become pregnant despite having frequent intercourse for more than one year.

INFERTILITY BELT. A geographic region stretching from east to west through central Africa. In some countries, one-third of all women are childless despite repeated attempts to have children. Although the medical problem is divided equally among men and women, it is typically women who are blamed for the infertility.

INHERITABLE GENETIC MODIFICATION (IGM). See GERMLINE GENE THERAPY.

INNER CELL MASS. The group of cells inside a mammalian embryo whose fate is to generate the entire embryo. The outer cells make up the trophoblast (q.v.), and they help form the placenta.

INTRACYTOPLASMIC SPERM INJECTION (ICSI). An in vitro fertilization procedure wherein a single sperm is injected into a mature egg. This procedure is used in cases in which the male partner produces too few sperm or in which the immune system of the female partner's reproductive tract destroys sperm.

IPSC. See INDUCED PLURIPOTENT STEM CELL.

IVF. See IN VITRO FERTILIZATION.

KARMA. In Hindu and Buddhist traditions, the idea that a person's actions in this and previous lives will determine the social fate of the embryo. Karma states that the actions performed in one's life will have a direct effect on one's children.

LABOR. Childbirth, involving contractions of the uterus, the thinning of the cervix, the pushing of the baby out of the birth canal, and the delivery of the placenta.

LESCH-NYHAN SYNDROME. A rare hereditary disease consisting of a self-mutilation syndrome, intellectual disability, and early death. It is linked to a gene on the X chromosome and therefore is expressed only in males.

LGBT (LESBIAN, GAY, BISEXUAL, TRANSGENDER). An abbreviation attempting to cover individuals whose sexual behaviors and preferences differ from those expected from their physical anatomy. An identification that does not correspond to the socially normal binary of male and female identities.

LH. See LUTEINIZING HORMONE.

LUTEINIZING HORMONE (LH). A mammalian pituitary hormone that stimulates estrogen production in the ovaries and testosterone production in the testes. A surge of LH in adult women prepares the oocyte for ovulation (q.v.).

MEIOSIS. The process of cell division by which a germ cell (q.v.) halves the number of chromosomes in its nucleus. The replication of DNA (making four copies of each gene) is followed by two divisions that make four cells, each having one copy. Most cells of the body have two copies of each gene. At fertilization, the sperm and egg fuse, and their nuclei come together, giving the normal two copies of each gene.

MENSTRUAL CYCLE. The hormonally induced rhythm whereby the reproductive areas of a woman's bodies are coordinated for reproduction. The eggs mature in the ovaries, the uterus prepares for receiving an embryo, and the mucus of the cervix prepares to allow sperm transport. If fertilization does not occur, the uterine lining is shed, and the cycle repeats until menopause, when no further eggs mature.

MESODERM. The central of the three layers that generate the embryo. It forms the heart, blood, kidneys, gonads, bones, and muscles.

METAPHOR. A figure of speech, using a direct means (not using "like or "as") to indicate similarity.

METHYL GROUP. The smallest organic chemical residue (a carbon atom with three hydrogen atoms). When placed onto DNA or the proteins surrounding the DNA, it can regulate the activity of that gene. Usually, placing methyl groups onto DNA or proteins prevents that gene from being active, whereas taking methyl groups off the DNA makes it active.

MIFEPRISTONE. Also called "RU486." An abortion-causing drug that is most effective in the embryonic period (less than seven weeks since the start of the last menstrual cycle). It blocks the activities of progesterone and causes the uterus to begin contractions.

MISCARRIAGE. Sometimes called "pregnancy loss" or "spontaneous abortion." A pregnancy that ends on its own, before it is viable. It is estimated that 10 to 25 percent of clinically recognized pregnancies end in miscarriage.

MISMATCH HYPOTHESIS. The idea that the food resources experienced by the fetus while in the uterus set the parameters of expression of those genes encoding the proteins involved in metabolism. If the environment of the baby does not match the environment of the fetus, health problems may ensue.

MITOCHONDRIA. Membrane-bound "organelles" (little organs) found in the cytoplasm (not the nucleus) of each cell. These structures use oxygen and food to generate energy. Human mitochondria also contain about three dozen genes (whereas the chromosomes of the nucleus have over twenty thousand genes).

MITOCHONDRIAL REPLACEMENT THERAPY (MRT). When mitochondrial genes are deficient, certain disease can develop. Since our mitochondria are derived from the egg and not from the sperm, one can place the mother's egg nucleus into an enucleated egg (q.v.) containing normal mitochondria and then perform in vitro fertilization using that egg. Or, one can place the newly formed zygote nucleus into another woman's egg.

MORNING-AFTER PILL. Also referred to as "emergency contraception" or the brand name "Plan B." A contraceptive pill that is nearly 90 percent protective against pregnancy if taken within seventy-two hours of unprotected intercourse. It prevents ovulation from occurring and therefore prevents fertilization. It does not cause abortion.

MORPHOGENESIS. The organization of cells during the development of the embryo into functional tissues and organs. This occurs through cell growth, cell migration, and cell death.

MULTIPOTENT STEM CELL. These are stem cells, like heart stem cells or blood stem cells, which can generate a subset of cell types. (Compare with pluripotent and totipotent stem cells.)

NARRATIVE. An account that ties together events. A story.

OOCYTE. A developing egg. In humans, oocytes are still in the ovary.

ORGANOGENESIS. The reciprocal interactions between tissues so that they form organs.

OVARIAN HYPERSTIMULATION SYNDROME. A spectrum of diseases resulting from the administration of hormone medications used to hyperstimulate the ovary to mature several oocytes simultaneously. The ovaries may become swollen and painful, and severe forms of this condition mandate hospitalization for loss of breath and vomiting.

OVARIES. The female gonads. The germ cells of the ovaries become the eggs.

OVIDUCTS. The Fallopian tubes. The tubular tissues leading from the uterus (womb) to each ovary. Capacitation (q.v.) of the sperm takes place in the oviduct, as does fertilization.

OVULATION. The release of an egg from an ovary. The egg is surrounded by the zona pellucida (q.v.) and the cumulus.

OVUM. Plural "ova." A mature egg. It is the ovum that joins with the sperm and is fertilized.

PLACENTA. The organ made by both the embryo and the mother. It serves to support the fetus structurally, allows the apposition of blood vessels so that nutrients and oxygen can enter the fetus while urea and carbon dioxide leave it, and downregulates the immune system so that the fetus is not rejected.

PLAN B. See MORNING-AFTER PILL.

PLEIOTROPY. The phenomenon in which one gene (or one protein) plays different roles in different cell types.

PLURIPOTENT STEM CELL. A cell that can generate all the different cell types of the embryonic body. Upon cell division, one cell remains a stem cell, whereas the other cell can differentiate according to its tissue environment. The major pluripotent cells are the embryonic stem cells of the inner cell mass. Recently, pluripotent cells have been induced by activating particular genes found in any body cell (q.v. induced pluripotent stem cell).

POLAR BODY. In female meiosis, the smaller cell, containing very little cytoplasm, is generated by the asymmetric division of the oocyte. These cells are not used for development.

POLYCYSTIC OVARY SYNDROME. A relatively common hormonal condition characterized by enlarged ovaries. It is usually caused when the ovaries produce too much testosterone. This inhibits ovulation and can be a source of infertility.

POSTMENOPAUSAL PREGNANCY. A pregnancy in a woman who is no longer ovulating. This usually involves in vitro fertilization with donor eggs.

PREGNANCY. Although colloquially indicating the time that a conceptus (q.v.) is developing within a woman, pregnancy is medically defined as beginning once the embryo has implanted into the uterus (about a week after fertilization).

PREIMPLANTATION GENETICS. Testing for genetic diseases or sex by taking a cell from an early embryo produced by in vitro fertilization (q.v.) and identifying its genes before implanting the embryo into the uterus.

PROGESTERONE. A steroid hormone made by the ovary that is critical for maintaining a mammalian pregnancy. It is also produced by the corona radiata (q.v.), and it may serve as a chemical attractant for the sperm.

PROTEIN PHARMACEUTICALS. Proteins that are found normally in most people that need to be replaced in people with certain diseases. Insulin, for instance, is a protein needed by people with diabetes, and clotting factors are proteins needed by people with hemophilia.

RECIPROCAL INDUCTION. A common mechanism of organogenesis (q.v.) whereby the chemicals from one tissue cause changes in a second tissue. As the second tissue changes, it makes chemicals that cause changes in the first tissue.

REPRODUCTIVE TOURISM. Also called "fertility tourism," "reprotravel," and "cross-border reproductive care," reproductive tourism involves traveling to another country whose laws allow certain ART procedures not permitted in one's own country (such as surrogacy).

REPROTRAVEL. See REPRODUCTIVE TOURISM.

RIBONUCLEIC ACID (RNA). The immediate product of DNA, the sequence of RNA reflects the sequence of chemicals in DNA. RNA goes from the nucleus into the cytoplasm. There, the sequence of chemicals in the RNA instructs the production of particular proteins.

RNA. See RIBONUCLEIC ACID.

RU486. See MIFEPRISTONE.

SIMILE. An explicit comparison (using "like" or "as") in which two unlike things are found to have some property in common. (Compare with "metaphor" and "analogy.")

SINGLE NUCLEOTIDE POLYMORPHISM (SNP). A variation in a single base pair of a DNA sequence. Most are harmless, and they are the most common genetic difference between people. This is what is used in forensic analysis to identify the perpetrator of a crime.

SINGLETON. A single-birth infant (i.e., not born from a multiple pregnancy, such as twins or triplets).

SNP. See SINGLE NUCLEOTIDE POLYMORPHISM.

SOMATIC CELL GENE THERAPY. A somatic cell is a cell in the body that is not a germ cell. Somatic cell gene therapy is thus the correction of a gene mutation in a somatic cell.

SPERM. The male gamete, produced in the testes.

SPERM BANK. A facility for storing sperm (usually in a frozen state) for later use. In agriculture, sperm banks collect and store sperm from prize bulls or stallions. In humans, sperm is stored for later use by women seeking artificial insemination.

STEM CELL. A relatively undifferentiated cell that, upon division, produces (1) another stem cell; and (2) a second cell that can be induced by its surrounding cells to differentiate (q.v.). In this manner, there is always a population of stem cells.

SUPEROVULATION. The result of ovarian hyperstimulation. The maturation of several oocytes into eggs each cycle, rather than just a single oocyte.

SURROGATE MOTHER. A gestational carrier. A woman who carries the conceptus (q.v.) of another couple. In gestational (full) surrogacy, the surrogate's pregnancy results from the transfer of an embryo generated by in vitro fertilization. In traditional ("partial") surrogacy, the surrogate's egg is used, but that egg is fertilized by the sperm of the male partner of the couple desiring a child. Here, the resulting child is genetically related to the surrogate.

TALMUD. The collection of Jewish writings (finished in about AD 500) explicating Biblical passages. Used as the springboard for debates concerning how to live a righteous life in accordance with Jewish law in various societies.

TEST-TUBE BABY. Slang term for a person who is the product of in vitro fertilization.

TOTIPOTENT STEM CELL. A cell that can generate an entire embryo, as well as the embryonic portion of the placenta. These are the earliest products of cell division, including the cells of the eight-cell embryo. (Compare with "pluripotent stem cell" and "multipotent stem cell.")

TRANSGENE. DNA from some other source that is inserted into the nucleus of a cell and becomes incorporated into a chromosome. Viruses, for instance, can carry a gene from one cell to another, making it a transgene.

TROPHOBLAST. The outer layer of the blastocyst (q.v.), it becomes the chorion, the embryonic portion of the placenta (q.v.).

TYRANNY OF CHOICE. The idea that too much choice makes decision-making difficult and causes unhappiness in people who feel they must make the "best choice."

ULTRASOUND. A technique using sound waves (of a higher frequency than humans can hear) to visualize soft tissue. It can visualize a fetus with enough resolution to see if it is male or female, which has caused problems with people using ultrasound for sex selection. Several states make it mandatory for a woman seeking an abortion to see ultrasound photographs of her fetus beforehand.

UMBILICAL CORD. The flexible blood vessel–containing tissue that connects the conceptus (q.v.) to the placenta.

UTERUS. Womb. Responsive to hormones such as estrogen and progesterone, it provides the support and housing for the embryo and fetus. The lowest end, the cervix, connect to the vagina, while the upper part connects to the oviducts (q.v.).

VIRAL VECTOR. A mechanism for bringing transgenes into a cell. In gene therapy, viral vectors are used to bring functional genes into cells whose genes contain mutations that prevent their functioning.

WRONGFUL BIRTH. A legal action, allowed in some countries, in which the parents of a child with congenital anomalies sue their physician, claiming that the physician did not counsel them properly on the risk of giving birth to a child with such serious conditions.

ZONA PELLUCIDA. An extracellular coat of proteins that surround the mammalian egg. Secreted by the oocytes, the zona pellucida is critical in sperm–egg recognition and in preventing the early embryo from adhering to the oviducts rather than the uterus.

ZYGOTE. The product of fertilization, the "fertilized egg."

NOTES

2. STORIES OF INFERTILITY AND ITS CONQUEST: THE SISTERHOOD OF BLOODY MARY

1. ICSI is the abbreviation for intracytoplasmic sperm injection, in which a sperm is injected directly into the cytoplasm of an unfertilized egg. The technique and its implications will be presented and discussed later in this book. For now, all we need to know is that it is a procedure used to circumvent problems of male infertility.

2. In 2014 (the latest year for which there are published statistics), about 173,200 cycles of ART were performed in the United States in attempts to produce babies, of which 57,323 led to full-term pregnancies (CDC 2016). That's 33 percent, a little less than one successful outcome for every three tries.

3. After a long series of failed attempts at different clinics, some patients actually prefer to stay away from anything "scientific" as they keep trying to have babies.

4. In politically correct terms, we should no longer be saying "infertile," but rather "involuntarily childless." Other labels now used include "nonmother" and "childfree." Note that all these terms still imply that something is missing. Some terms have recently been introduced to attempt to fix these negative implications. The negative associations are primarily directed at women, for it is typically women, not men, who "choose not to have children." Single, childless men are often more favorably perceived as bon vivants, which has a quite charming connotation.

5. The expression "IVF treadmills" (note the plural, implying more than one cycle) has become commonplace in related literature since the late 1990s, with a vast number of publications using the term in their titles. You can find some interesting references to European and worldwide folklore–based treatments for infertility detailed in the introductory chapter to Joseph Needham's (1934) *A History of Embryology*.

6. One of the most notorious examples of this last pattern is the tragic and bloody story of Queen Mary Tudor of England. There is an enormous wealth of literature on this subject. See David Loades (2006) for a simple but reliable account.

7. Or not that unconscious. In *Not Trying* (2014), Kristin Wilson states that when she started looking for voluntarily childless women she quickly realized she was after "a hidden population." People generally hide for a reason.

8. You can find this refrain in the works of evolutionary psychologists such as Jerome Barkow, Leda Cosmides, and John Tobby. Most prominent are Edward O. Wilson, whose *Sociobiology* (1975) became a rallying standard for those believing that all biology was an epiphenomenon of our genes, and Richard Dawkins, whose *Blind Watchmaker* (1986) claimed that propagating our genes was by far the most important function of life.

9. Men often don't talk. This is the reason I spoke only with women. Their husbands, when they were there, just nodded. However, when our infertility helpline inaugurated a "Husband Happy Hour" with spicy mixed drinks, homemade pastries, and no partners present, the guys showed up and talked their hearts out to each other and then went to a bar after the official sessions closed.

10. Or just try, for Christ's sake.

11. We will be discussing this later in the book. However, the crisis is dire. To access an alarming picture of the living hell women submit themselves to in Northern Africa as they try to get pregnant in order not to lose their husbands or be ostracized by society, see Inhorn's (1994a) *Quest for Conception*. For a global picture of the suffering of infertile women, see Jimmy Carter's (2014) *A Call to Action*.

3. FERTILIZATION: TWO CELLS AT THE VERGE OF DEATH COOPERATE TO FORM A NEW BODY THAT LASTS DECADES

1. Much of the material in this chapter is detailed in Gilbert and Barresi (2016).

2. The word "usually" is used here because the X and Y chromosomes only *initiate* the process. There are numerous genes on many other chromosomes that become involved in gonad formation and whose absence or loss of function can prevent the chromosomal sex from being realized. Moreover, once they are formed, the testes and ovaries secrete hormones that tell the rest of the body whether to have a female or male character. As we will see later, mutations of genes involved in hormone production can also change the sex of the baby.

3. For a video on the IVF process, see Andrea Vidali's "In Vitro Fertilization: A Short Animated Review," at https://vimeo.com/22048103.

4. A period of twenty-eight days is about average. However, some women have shorter or longer cycles. Most women have cycles that vary a few days each time. Primates, including humans, appear to be the only animals that menstruate. The reasons for this are not fully known but may include a cleansing of bacteria from the uterus each month.

5. Human chorionic gonadotropin (hCG) is the chemical being measured by most home pregnancy kits. A positive test indicates that there is hCG in your urine, and therefore, some cells making hCG in your body. Since hCG is made only by the embryo (and

later by the placenta), it indicates pregnancy. However, during the first few days of development, the embryo hasn't started making this protein yet (it starts making it at the time it implants into the uterus), and levels in the urine become detectable about twelve days after fertilization. (It should be remembered that pregnancy begins at implantation. "Pregnancy tests" work at the time of implantation, not fertilization.)

6. Inside boxes of morning-after pills, there is an insert that claims that the pills may cause abortion. This is scientifically unfounded and the result of political lobbying by organizations (such as Hobby Lobby and the American Association of Pro-Life Obstetricians and Gynecologists) that equate contraception with abortion (Dreweke 2014; Sneed et al 2014). The web page on emergency contraception of the U.S. government's medical website, MedlinePlus (https://medlineplus.gov/ency /article/007014.htm), once made the same claim but no longer does so.

4. FERTILITY RITES: RITES: ARTIFICIAL INSEMINATION AND IN VITRO FERTILIZATION—THEIR HOPES AND THEIR FEARS

1. The Latin is a joy to read: *Fecerunt medici cannam auream, quam Regina in vulvam recepit, an per ipsam semen inicere posset; nequivit tamen. Mulgere item fecerunt feretrum eius, et exivit sperma, sed aquosum et sterile.* This topic is full of legends, many of which are difficult to verify. It is sometimes thought that London-based doctor John Hunter, nicknamed the "father of scientific surgery," might have tried his hand at human AI in the 1770s, trying to aid a cloth merchant with severe spasms that caused his semen to escape during coitus by collecting it in a warmed syringe and injecting the sample into his wife's vagina. Sims remains controversial because he performed many of his experimental surgeries on slave women in the American South. The concept of informed consent did not exist until after World War II.

2. For timelines of IVF and medical ethics milestones, see "Stem Cells Across the Curriculum" at www.stemcellcurriculum.org/timelines.html.

3. Duke University has been keeping an interesting blog on issues of molecular genetics data concerning gene transmission through several generations, and some sections are worth reading for a clear view of what we know by now—and what a mess our genes can be. Two further interesting articles on this topic are "Autosomal DNA, Ancient Ancestors, Ethnicity and the Dandelion" (http://dna-explained.com/2013/08/05/autosomal-dna -ancient-ancestors-ethnicity-and-the-dandelion/) and "How Much Of Your Genome Do You Inherit from a Particular Ancestor?" (http://gcbias.org/2013/11/04/how-much-of -your-genome-do-you-inherit-from-a-particular-ancestor/).

4. Lest someone think that this was a minor part of Pope John Paul II's talk at the Eleventh International Colloquium on Roman and Canon Law, the pope's speech was titled, loud and clear, "I Appeal to the World's Scientific Authorities: Halt the Production of Human Embryos!" The theme had already been touched on in the March 25, 1995, papal encyclical, *Evangelium Vitae.*

5. NORMAL DEVELOPMENT AND THE BEGINNING OF HUMAN LIFE: WHY SCIENTISTS ARE BEING ASKED THEOLOGICAL QUESTIONS AND WHY THEOLOGIANS ARE BEING ASKED SCIENTIFIC QUESTIONS

1. The material in this chapter is taken largely from Gilbert and Barresi (2016) and Carlson (2014).
2. Much of the material from this section is drawn from Gilbert et al. (2005) and Gilbert (2008).
3. There is an interesting resonance here with modern embryology. Until gastrulation, the embryo may be said to be "vegetative." It can grow, and the cells can be like those of plants, forming separate embryos if separated. Gastrulation, however, is the *sine qua non* of being an animal. It is movement. So here, the "animal" principle takes over. Last, when we develop our EEG pattern, we have the capacity to become "rational" beings. So the places where science has claimed individual human life begins correspond to these Aristotelian categories.
4. As a Southern European Catholic woman, Clara wants to state for the record that Catholics around the world differ on this issue. She was part of the movement in Portugal that fought a difficult and dangerous struggle for the legalization of abortion in her country after it was strictly forbidden during the many decades of a severe fascist dictatorship. The anti-abortion policy was still forced on her teenage high school friends who had to resort to the services of clandestine midwives on shady street corners.

6. TECHNOLOGICAL MOTHERHOOD

1. Because of the significant number of prostitutes acting as surrogate mothers, and of Philippine girls recruited underground explicitly for the task of surrogacy, several countries (mainly in Europe) have deemed "foster uteri" illegal.
2. In a completely serendipitous way, I happen to know this from my own personal experience: After my first failed IVF procedures, a Spanish doctor running an egg donation program tried to persuade me to try this method and gave me a colorful brochure presenting his clinic, his staff, his method, his donors, and also a chart of his prices with the clinic's different payment options. Both the donor and their family are checked, too.

7. CLONING ANIMALS, CELLS, AND GENES: WHERE DID CLONING COME FROM, AND WHERE IS IT GOING TO RIGHT NOW?

1. Much of the material in this chapter can be found in Gilbert (2014).
2. This topic will be discussed in chapter 8, which examines the "banking" of blood stem cells (see also Sibov et al. 2012). At present, the techniques for obtaining these cells are still experimental and costly, and only a small proportion of the transferred cells actually

become established in the host body (Roura et al 2016). The American College of Obstetricians and Gynecologists (2015) warns against the storage of umbilical cord blood as "biologic insurance" against future disease.

8. GLORY DAYS : MY PERSONAL ACCOUNT OF CLONING

1. A postdoctoral fellow is a limbo position, in which you have a PhD but need to show that you can do "independent research." So you work in the laboratory of an established scientist, and that scientist pays your salary. It's a heady time, as you can focus your entire being on doing research without the responsibilities of teaching, grant writing, or sitting on committees.

2. Or, more cynically, "to be useful for something that later on will make somebody rich and famous while saving mankind."

3. We now know that the methylation of DNA, a process that regulates the activity of genes, is often screwed up in clones. This probably accounts both for the lack of success in obtaining cloned animals and the poor health of those that survive. More about this later.

4. Yes, Scott had to call the publishers of his textbook and tell them to stop the manuscript from going to the printer. He then had to remove the sentences explaining that no mammal had been cloned from an adult somatic cell and provide them with exactly the same number of letters saying that the feat had been accomplished.

5. For the incredible story of these cells, derived from the cancerous cells of Henrietta Lacks's cervix, see Skloot (2010). By now it is estimated that scientists have grown some twenty tons of her cells, and there are almost eleven thousand patents involving her cells. The Lacks family, however, lives below the poverty line and often lacks health insurance.

6. Indeed, the story goes that after arriving home that evening, Molina's wife asked him how his day had been. "The day was fine," Molina said, "but the world is going to come to an end."

7. Although this line of research ended up proving quite fruitless within the confines of mammalian cloning, its initial rationale was the production of human products within the eggs of other species.

8. Now, we'd probably just give him a copy of this book.

9. I'm still thankful to this day for each and every one of the many brave people in those audiences who made it a point to stand up and ask excellent questions. I'm thankful to the several other guests at the speakers' table who told those chairs they had nothing to say after all, because they wished they had been the ones knowing enough about stem cells to give my talk themselves. Since these situations generally ended up with the chair leaving the room and the stem cell conversation continuing, I should also thank the armed members of the local police who showed up after a while, having certainly been told something dangerous was happening, and instead of forcing us to leave, just stood by and listened until closing time. People present at these high-adrenaline stem cell "rebellions" were often so excited afterward that they wanted to take me out for a very late dinner,

a great local bar, or somebody's house for a party, so I should also apologize for always excusing myself so that I could go back to my hotel and crash. Those days were tough.

10. For instance, we don't know how long stem cells will remain viable in liquid nitrogen, because the experiment has never been done before. Moreover, not *all* cells in the umbilical cord are pluripotent stem cells. Some stem cells are more plastic than others, and some are already committed to certain fixed fates. In some cases (as in autoimmune diseases), there is a pernicious environment that won't allow certain cells to live. There's no reason to suspect that the new cells will live while the old cells didn't. Worse, stem cells might form tumors. We cannot do experiments on humans as we can on mice, so we have to extrapolate from the results of one organism to another. This is always risky.

11. Being a jurist means having completed law school but not yet having taken the bar exams.

9. INFERTILITY WARS: HOW LIFE FEELS AFTER EVERYTHING FAILS, AND, BY THE WAY, HOW DO WE SURVIVE IT?

1. Twice I had to fight my way through this sort of hostility to manage getting my talks to their point on time, and I'm not at all into fighting. I was always exhausted by the end. A while later, I quit attending those meetings and started turning down invitations. I would like to believe things have changed.

2. The list is long, and for the most, the titles are sad. For a thoroughly informing selection of essays on this subject, see *Men, Women, and Infertility* (Zolbrod 1993). Starting with the sobering thought, "Infertility is an equal-opportunity crisis, touching people from urban, suburban, and rural areas, from all sexual orientations and from every ethnic, racial, class, and religious background," it lays down an excellent perspective of what life is like for those 12 percent of American couples of childbearing age out there who cannot conceive when they wish to. For well-tempered personal stories, see Renate D. Klein's (1989) *Infertility: Women Speak Out About Their Experiences of Reproductive Medicine*. The aggressively titled *My Body—My Decision!* (Curtis et al. 1986) is for those seeking information on different types of birth control and ART. Now, from the male perspective, we can start with classic case study medical reports in the soberly titled *Impotence* (Wagner 1981) or turn to the vivid psychological account of personal struggles in *Male Infertility: Men Talking* (Mason 1993). For hardcore social science studies see, for instance, Linda Hammer Burns's *Psychological Changes in Infertility Patients* (2005) or Marcia C. Inhorn and Frank van Balen's *Infertility Around the Globe* (2002).

3. Guys, it might not have been your fault, but honestly—are you sure you really know what a worst-of-all-nightmares this lifestyle represents for a woman?

4. It should be considered relevant that by now this constitutes an entire subspecialty with its own meetings and history (Boivin and Gameiro 2015).

5. I can laugh at it *now*. It felt really terrible twenty years ago.

6. My psychologist colleague by then had a patient trying to raise triplets with incurable limitations, born from IVF after five failed cycles.

7. This is a disorder in which the endometrium, the lining of the uterus that is shed each month and which can bind to the early embryo, grows outside the uterus. Endometriosis can cause moderate to severe pain. It can also cause, as it probably did in my case, infertility.

8. And this information comes after checking through what was available at Harvard; the University of Massachusetts, Amherst; Amherst College; Smith College; and Mount Holyoke College—places where there should be plenty of people interested in this topic!

9. You have no idea. Kim Kardashian's failed IVF. Even *that* exists now.

10. Research financed by Microsoft, published in the summer of 2014, shows that people tend to turn to something else if the download of a site takes more than the blink of an eye.

11. That's right, *pain*! Take that, you family therapist at my dinner party, whose clients I've pitied ever since.

12. These are rough estimates taken directly from a number of modern infertility clinics. Results published in journals or presented at meetings tend to be more optimistic. Makes you wonder why.

13. These estimates are really *averages* taken from the web. Moreover, different countries regulate their clinics differently, which makes data difficult to compare. This is also, of course, for legally provided services.

14. Women at these globalized IVF clinics often say, "This is a lot more than tourism; it's desperation," and they often are there by themselves.

15. At first, this trend involved only European countries, mainly Belgium, the Czech Republic, Denmark, Slovenia, Spain, and Switzerland. The first study on the subject revealed that there were already up to thirty thousand cross-border IVF cycles being performed in Europe each year, involving up to fourteen thousand patients.

16. As a prospective Muslim client put it in 2015, "The moment you type in the word 'infertility,' you just open up 'India.'"

10. THE HUMAN CONDITION OF FEAR AND WONDER: IN CELEBRATION OF BODIES

1. The Dalai Lama is opposed to abortion, an opposition grounded in his belief that all animal life is "inestimably precious" and that violence should not be perpetrated on any animal. However, he continues, abortion could be done if the birth of the child would cause "severe suffering for certain members of the family" (Dalai Lama 1996, 43).

REFERENCES

Adams, D. 2002. *The Salmon of Doubt: Hitchhiking the Galaxy One Last Time*. New York: Harmony.

Águeda Maujo, H., and C. Pinto-Correia. 2004. *A Dor Secreta da Infertilidade: História de Uma Mulher que não Pode ter Filhos*. Lisbon: Presença.

Albertus Magnus, n.d., ca. 1249. Quoted in Demaitre, L., and A. A. Travill. 1980. "Human Embryology and Development in the Works of Albertus Magnus." In *Albertus Magnus and the Sciences: Commemorative Essays*, edited by J. A. Weisheipl. Toronto: Pontifical Institute of Mediaeval Studies.

Alden, P. B. 1996. *Crossing the Moon: A Journey Through Infertility*. Saint Paul, MN: Hungry Mind.

American College of Obstetricians and Gynecologists (ACOG). 2015. "ACOG Committee Opinion No. 648: Umbilical Cord Blood Banking." *Obstet Gynecol* 126 (6): e127–e129.

American Society for Reproductive Medicine (ASRM). 2008. "Ovarian Hyperstimulation Syndrome." *Fertility and Sterility* 90: S188–193.

——. 2014. "Can I Freeze My Eggs to Use Later if I'm Not Sick?" http://reproductivefacts.org/FACTSHEET_Can_I_freeze_my_eggs_to_use_later_if_Im_not_sick/.

Anderson, R. E. 2004. "Ethics of Embryonic Stem Cells." *New England Journal of Medicine* 351 (16): 1687–1690.

Andrews, L. B. 1999. *The Clone Age: Adventures in the New World of Reproductive Technology*. New York: Henry Holt.

Angier, N. 1992. "A First Step in Putting Genes into Action: Bend the DNA." *New York Times*, August 4. www.nytimes.com/1992/08/04/science/a-first-step-in-putting-genes-into-action-bend-the-dna.html?pagewanted=all.

——. 1999a. *Woman: An Intimate Geography*. New York: Houghton Mifflin.

——. 1999b. "Baby in a Box." *New York Times Magazine*. https://partners.nytimes.com/library/magazine/millennium/m2/angier.html.

Aquinas, T., n.d., ca. 1260. *Commentary on the Sentences III*, dist. 3, question 5; *Summa Contra Gentiles II*, chapter 89. Translated by J. Kenny. New York: Hanover. http://dhspriory.org/thomas/ContraGentiles.htm.

Aristotle. 350 BCE. *Metaphysics* 12, 982b12.

Augustine of Hippo, n.d., ca. 410. *City of God*, 21.8.

Austin, C. R. 1952. "The 'Capacitation' of Mammalian Sperm." *Nature* 170 (4321): 326.

Australian Cerebral Palsy Register Group (ACPRG). 2013. *Australian Cerebral Palsy Register Report 2013*. Allambie Heights, New South Wales, Australia: Cerebral Palsy Alliance.

Bacon, F. 1605. *The Advancement of Learning*. Book I, Chapter 3.

Balter, M. 2014. "Human Altruism Traces Back to the Origins of Humanity." *Science*, August 27. www.sciencemag.org/news/2014/08/human-altruism-traces-back-origins-humanity.

Baltimore, D., P. Berg, M. Botchan, D. Carroll, R. A. Charo, G. Church, J. E. Corn, G. Q. Daley, J. A. Doudna, M. Fenner, H. T. Greely, M. Jinek, G. S. Martin, E. Penhoet, J. Puck, S. H. Sternberg, J. S. Weissman, and K. R. Yamamoto. 2015. "A Prudent Path Forward for Genomic Engineering and Germline Modification." *Science* 348 (6230): 36–38.

Barbour, I. 1971. *Issues in Science and Religion*. New York: Harper.

Barker, D. J. P. 1989. "Rise and Fall of Western Diseases." *Nature* 338 (6214): 371–372.

——. 1994. *Mothers, Babies and Disease in Later Life*. London: Churchill Livingstone.

——. 1995. "Fetal Origins of Coronary Heart Disease." *British Medical Journal* 311 (6998): 171–174.

Balayla, J., O. Sheehy, W. D. Fraser, et al. 2017. "Neurodevelopmental Outcomes After Assisted Reproductive Technologies." *Obstetrics and Gynecology*. doi:10.1097/AOG.0000000000001837.

Bavister, B. D. 2002. "A History of In Vitro Fertilization." *Reproduction* 124: 181–196.

Belva, F., M. Bonduelle, M. Roelants, et al. 2016. "Semen Quality of Young Adult ICSI Offspring: the First Results." *Human Reproduction* 31 (12):2811–2820.

Berend, Z. 2010. "Surrogate Losses: Understanding of Pregnancy Loss and Assisted Reproduction Among Surrogate Mothers." *Medical Anthropology Quarterly* 24 (2): 240–262.

Bhattacharya, S., and A. Templeton. 2000. "In Treating Infertility, Are Multiple Pregnancies Unavoidable?" *New England Journal of Medicine* 343 (1): 58–60.

Bialosky, J., and H. Schulman, eds. 1999. *Wanting a Child*. New York: Farrar, Straus and Giroux.

Biology and Gender Study Group (BGSG), A. Beldecos, S. Bailey, S. Gilbert, K. Hicks, L. Kenschaft, N. Niemczyk, R. Rosenberg, S. Schaertel, and A. Wedel. 1988. "The Importance of Feminist Critique for Contemporary Cell Biology." *Hypatia* 3 (1): 61–76.

Boerma, J. T., and Z. Mgalla, eds. 2001. *Women and Infertility in Sub-Saharan Africa: A Multi-Disciplinary Perspective*. Amsterdam: Royal Tropical Institute.

Boggs, B. 2016. *The Art of Waiting: On Fertility, Medicine, and Motherhood*. Minneapolis: Graywolf.

Boiani, M., S. Eckardt, H. R. Schöler, and K. J. McLaughlin. 2002. "Oct4 Distribution and Level in Mouse Clones: Consequences for Pluripotency." *Genes and Development*. 16 (10): 1209–1219.

Boivin, J., and S. Gameiro. "Evolution of Psychology and Counseling in Infertility." *Fertility and Sterility* 104 (2): 251–259.

Bonner, G. 1985. "Abortion and Early Christian Thought." In *Abortion and the Sanctity of Human Life*, edited by J. H. Channer, 93–122. Exeter: Paternoster.

Briggs, R., and T. J. King. 1952. "Transplantation of Living Nuclei from Blastula Cells into Enucleated Frogs' Eggs." *Proceedings of the National Academy of Sciences of the United States of America* 38 (5): 455–463.

Brüstle O., K. N. Jones, R. D. Learish, K. Karram, K. Choudhary, O. D. Wiestler, I. D. Duncan, and R. D. McKay.1999. "Embryonic Stem Cell-Derived Glial Precursors: A Source of Myelinating Transplants." *Science* 285 (5428): 754–756.

Buklijas, T., and N. Hopwood. 2010. "The Lonesome Space Traveller." *Making Visible Embryos*. www.hps.cam.ac.uk/visibleembryos/s7_4.html.

Burdge, G. C., J. Slater-Jefferies, C. Torrens, E. S. Phillips, M. A. Hanson, and K. A. Lillycrop. 2006. "Dietary Protein Restriction of Pregnant Rats in the F_0 Generation Induces Altered Methylation of Hepatic Gene Promoters in the Adult Male Offspring in the F_1 and F_2 Generations." *British Journal of Nutrition* 97 (3): 435–439.

Burgstaller, J. P. and G. Brem. 2016. "Aging of Cloned Animals." *Gerontology*. doi:10.1159/000452444.

Burkart, J. M., O. Allon, F. Amici, C. Fichtel, C. Finkenwirth, A. Heschl, J. Huber, K. Isler, Z. K. Kosonen, E. Martins, E. J. Meulman, R. Richiger, K. Rueth, B. Spillmann, S. Wiesendanger, and C. P. van Schaik. 2014. "The Evolutionary Origin of Human Hyper-Cooperation." *Nature Communications* 5: 4747. doi:10.1038/ncomms5747.

Burns, L. H. 2005. *"Psychological Changes in Infertility Patients."* In *Frozen Dreams: Psychodynamic Dimensions of Infertility and Assisted Reproduction*, edited by A. Rosen and J. Rosen, 3–29. Hillsdale, NJ: Analytic.

Buss, M. 1967. "The Beginning of Human Life as an Ethical Problem." *Journal of Religion* 47 (3): 244–255.

Cangiamila, F. E. 1758. *Embryologia Sacra*. Palermo: Francesco Valenza.

Caplan, A. 2005. "Fertility Clinics Vary Widely on Who Gets Treatment." *CNN Health*. www.cnn.com/2005/HEALTH/01/19/fertility.ethics.ap/; http://www.ncpa.org/sub/dpd/index.php?Article_ID=1175

Carlson, B. M. 2014. *Human Embryology and Developmental Biology*. 5th ed. Philadelphia: Elsevier.

Carlson, E. A. 2001. *The Unfit: A History of a Bad Idea*. Cold Spring Harbor, MA: Cold Spring Harbor Laboratory Press.

Carmichael, M. 2004. "No Girls, Please." *Newsweek*, January 143 (4): 50. At http://www.newsweek.com/no-girls-please-125881.

Carter, J. 2014. *A Call to Action: Women, Religion, Violence, and Power*. New York: Simon & Schuster.

Caserta D., G. Bordi, F. Ciardo, et al. 2013. "The Influence of Endocrine Disruptors in a Selected Population of Infertile Women." *Gynecology and Endocrinology* 29: 444–447.

Center for Genetics and Society (CGS). 2016a. "About Inheritable Genetic Modification." www.geneticsandsociety.org/section.php?id=108.

——. 2016b. "3-Person IVF: A Resource Page." http://www.geneticsandsociety.org/article.php?id=6527.

Centers for Disease Control and Prevention (CDC). 2012. "2012 Assisted Reproductive Technology National Summary Report ART Success Rates." http://www.cdc.gov/art/reports/2012/national-summary.html.

——. 2014. "Infertility." Last updated July 15, 2016. www.cdc.gov/nchs/fastats/infertility.htm.

——. 2016. "ART Success Rates." www.cdc.gov/art/artdata/index.html.

Chabbert-Buffet, N., G. Meduri, P. Bouchard, and I. M. Spitz. 2005. "Selective Progesterone Receptor Modulators and Progesterone Antagonists: Mechanisms of Action and Clinical Applications." *Human Reproduction Update* 11 (3): 293–307.

Champagne, F. A., I. C. Weaver, J. Diorio, S. Dymov, M. Szyf, and M. J. Meaney. 2006. "Maternal Care Associated with Methylation of the Estrogen Receptor-alpha1b Promoter and Estrogen Receptor-alpha Expression in the Medial Preoptic Area of Female Offspring." *Endocrinology* 147 (6): 2909–2915.

Chang, M. C. 1951. "Fertilizing Capacity of Spermatozoa Deposited into the Fallopian Tubes." *Nature* 168 (4277): 697–698.

Chapman, A. R., and M. S. Frankel. 2003. "Framing the Issues." In *Designing Our Descendents: The Promises and Perils of Genetic Modifications*, edited by A. R. Chapman and M. S. Frankel, 3–19. Baltimore: Johns Hopkins University Press.

Chavez, S. L., K. E. Loewke, J. Han, F. Moussavi, P. Colls, S. Munne, B. Behr, and R. A. Reijo Pera. 2012. "Dynamic Blastomere Behaviour Reflects Human Embryo Ploidy by the Four-Cell Stage." *Nature Communications* 3: 1251.

Chen, X., M. Chen, B. Xu, R. Tang, X. Han, Y. Qin, B. Xu, B. Hang, Z. Mao, W. Huo, Y. Xia, Z. Xu, and X. Wang. 2013. "Parental Phenols Exposure and Spontaneous Abortion in Chinese Population Residing in the Middle and Lower Reaches of the Yangtze River." *Chemosphere* 93 (2): 217–222.

Childs, B. 2003. *Genetic Medicine: A Logic of Disease*. Baltimore: Johns Hopkins University Press, 108.

Cohen, I. G., G. Q. Daley, and B. Y. Adashi. 2017. "Disruptive Reproductive Technologies." *Science Translational Medicine* 9 (372) doi:10.1126/scitranslmed.aag2959.

Cohen-Dayag, A., I. Tur-Kaspa, J. Dor, S. Mashiach, and M. Eisenbach. 1995. "Sperm Capacitation in Humans Is Transient and Correlates with Chemotactic Responsiveness to Follicular Factors." *Proceedings of the National Academy of Sciences of the United States of America* 92 (24): 11039–11043.

Congregation for the Doctrine of the Faith (CDF). 1987. *Donum Vitae* [Respect for Human Life]. Vatican City: Congregation for the Doctrine of the Faith.

Couzin-Frankel, J. 2015. "Eggs Unlimited." *Science* 350 (6261): 620–624.

Cox, D. B., R. J. Platt, and F. Zhang. 2015. "Therapeutic Genome Editing: Prospects and Challenges." *Nature Medicine* 21 (2): 121–131.

Crichton, M. 1990. *Jurassic Park*. New York: Alfred A. Knopf.

Curtis, L. R., G. B. Curtis, and M. K. Beard. 1986. *My Body—My Decision! What You Should Know About the Most Common Female Surgeries*. Tucson: Body.

Cusk, R. 2016. "What She Bears," *New York Times Book Review*, September 4. Dali Lama. 1993. "New York Times Interview with the Dalai Lama." By C. Dreifus. *New York Times*, November 28. www.sacred-texts.com/bud/tib/nytimes.htm.

Dalai Lama. 1995. *The World of Tibetan Buddhism*. Boston: Wisdom Publications.

——. 1996. *Beyond Dogma: Dialogues and Discourses*. Berkeley, CA: North Atlantic.

Das, K. (1929) 1993. *Obstetric Forceps: Its History and Evolution*. Leeds: Medical Museum Publishing.

DasGupta, S., and S. DasGupta, eds. 2014. *Globalization and Transnational Surrogacy in India: Outsourcing Life*. New York: Lexington.

Dawkins, R. 1976. *The Selfish Gene*. Oxford: Oxford University Press.

——. 1986. *The Blind Watchmaker*. New York: Norton.

DeMarco, D. 1984. "The Roman Catholic Church and Abortion: A Historical Perspective." *Homiletic Press and Pastoral Review*, July 1984, 59–66. www.catholicculture.org/culture/library/view.cfm?id=3361.

Den Boer, A., and V. M. Hudson. 2002. "A Surplus of Men, a Deficit of Peace: Security and Sex Ratios in Asia's Largest States." *International Security* 26 (4): 5–38.

Desmond, A. 1997. *Huxley: From Devil's Disciple to Evolution's High Priest*. Reading, MA: Addison-Wesley.

Diamond, G. 1994. *Full Circle*. New York: Feldheim.

Dobzhansky, T. 1976. "Living with the Biological Revolution." In *Man and the Biological Revolution*, edited by T. Dobzhansky and R. H. Haynes. 21–45. Toronto: York University Press.

Domar, A. D., K. L. Rooney, M. Milstein, and L. Conboy. 2015. "Lifestyle Habits of 12,800 IVF Patients: Prevalence of Negative Lifestyle Behaviors, and Impact of Religion and Insurance Coverage." *Human Fertility* 18 (4): 253–257.

Dreweke, J. 2014. "Contraception Is Not Abortion: The Strategic Campaign of Antiabortion Groups to Persuade the Public Otherwise." *Guttmacher Policy Review* 17 (4). https://www.guttmacher.org/gpr/2014/12/contraception-not-abortion-strategic-campaign-antiabortion-groups-persuade-public.

Ducibella, T., and R. Fissore. 2008. "The Roles of Ca^{2+}, Downstream Protein Kinases, and Oscillatory Signaling in Regulating Fertilization and the Activation of Development." *Developmental Biology* 315 (2): 257–279.

Dworkin, A. 1983. *Right-Wing Women: The Politics of Domesticated Females*. London: Women's Press.

Edward, D. A., P. Stockley, and D. J. Hosken. 2014. "Sexual Conflict and Sperm Competition." *Cold Spring Harbor Perspectives in Biology* 7 (4): a017707.

Erb, B. J. 1999. "Deconstructing the Human Egg: The FDA's Regulation of Scientifically Created Babies." *Roger Williams University Law Review* 5 (1): 273–313.

Eugenics Archive. 2016. www.eugenicsarchive.org/eugenics/.

Evangelical Lutheran Church in America (ELCA). 1991. "Social Statement on Abortion" https://www.elca.org/Faith/Faith-and-Society/Social-Statements/Abortion.

Evans, M. J., and M. H. Kaufman. 1981. "Establishment in Culture of Pluripotent Cells from Mouse Embryos." *Nature* 292 (5819): 154–156.

Falk, M. J., A. Decherney, and J. P. Kahn. 2016. "Mitochondrial Replacement Technique: Implications for the Clinical Community." *New England Journal of Medicine* 374 (12): 1103–1106.

Feil, R., and M. F. Fraga. 2012. "Epigenetics and the Environment: Emerging Patterns and Implications." *Nature Reviews Genetics* 13: 97–109.

Fiorina, C. 2015. Quoted in N. Nazworth, "Interview: Potential GOP Presidential Candidate Carly Fiorina Talks Abortion, Common Core, Gay Marriage and Her Christian Faith." *Christian Post*, February 6. www.christianpost.com/news/interview -potential-gop-presidential-candidate-carly-fiorina-talks-abortion-common-core -gay-marriage-and-her-christian-faith-133652/#jRPJLB81tV6hewWT.99www .christianpost.com/news/interview-potential-gop-presidential-candidate-carly -fiorina-talks-abortion-common-core-gay-marriage-and-her-christian-faith-133652/.

Fleming, T. P. 1987. "A Quantitative Analysis of Cell Allocation to Trophectoderm and Inner Cell Mass in the Mouse Blastocyst." *Developmental Biology* 119 (2): 520–531.

Flower, M. J. 1985. "Neuromaturation of the Human Fetus." *Journal of Medical Philosophy* 10 (3): 237–251.

Ford, N. M. 1988. *When Did I Begin? Conception of the Human Individual in History.* New York: Cambridge University Press.

Fraga, M. F., E. Ballestar, M. F. Paz, S. Ropero, F. Setien, M. L. Ballestar, D. Heine-Suñer, J. C. Cigudosa, M. Urioste, J. Benitez, M. Boix-Chornet, A. Sanchez-Aguilera, C. Ling, E. Carlsson, P. Poulsen, A. Vaag, Z. Stephan, T. D. Spector, Y. Z. Wu, C. Plass, and M. Esteller. 2005. "Epigenetic Differences Arise During the Lifetime of Monozygotic Twins." *Proceedings of the National Academy of Sciences of the United States of America* 102 (30): 10604–10609.

Freedman, N. 1973. *Joshua, Son of None.* New York: Delacorte.

Frenkel, D. A. 2001. "Legal Regulation of Surrogate Motherhood in Israel." *Medicine and Law* 20 (4): 605–612.

Friedmann, T., E. C. Jonlin, N. M. P. King, B. E. Torbett, N. A. Wivel, Y. Kaneda, and M. Sadelain. 2015. "ASGCT and JSGT Joint Position Statement on Human Genomic Editing." *Molecular Therapy* 23: 1282.

Fritz, M. A., and L. Speroff. 2010. *Clinical Gynecological Endocrinology.* Philadelphia: Williams and Wilkins.

Fritz, R., C. Jain, and D. R. Armant. 2014. "Cell Signaling in Trophoblast–Uterine Communication." *International Journal of Developmental Biology* 58 (2–4): 261–271.

Gilbert, S. F. 1979. "The Metaphorical Structuring of Social Perceptions." *Soundings* 62:166–186.

——. 2002. "Genetic Determinism: The Battle Between Scientific Data and Social Image in Contemporary Developmental Biology." In *On Human Nature: Anthropological, Biological, and Philosophical Foundations*, edited by A. Grunwald, M. Gutmann, and E. M. Neumann-Held, 121–140. New York: Springer.

——. 2008. "When 'Personhood' Begins in the Embryo: Avoiding a Syllabus of Errors." *Birth Defects Research Part C: Embryo Today: Reviews* 84 (2): 164–173.

——. 2013. "Wonder and the Necessary Alliances of Science and Religion." *Euresis Journal* 4: 7–30.

——. 2014. *Developmental Biology*. 10th ed. Sunderland, MA: Sinauer.

——. 2015a. "DNA as Our Soul: Don't Believe the Advertising." *Huffington Post*, November 18. www.huffingtonpost.com/scott-f-gilbert/dna-as-our-soul-believing _b_8590902.html.

——. 2015b. "Republicans Need to Be Countered on False Claims About Embryos." *Huffington Post*, September 23. www.huffingtonpost.com/scott-f-gilbert/countering -republican-claims-embryos_b_8152028.html.

Gilbert, S. F., and M. Barresi. 2016. *Developmental Biology*. 11th ed. Sunderland, MA: Sinauer.

Gilbert, S. F., and S. Braukmann. 2011. "Fertilization Narratives in the Art of Gustav Klimt, Diego Rivera, and Frida Kahlo: Repression, Domination, and Eros among Cells." *Leonardo* 44: 221–227.

Gilbert, S. F., and D. Epel D. 2015. *Ecological Developmental Biology : The Developmental Integration of Evolution, Development, and Medicine*. 2nd ed. Sunderland, MA: Sinauer.

Gilbert, S. F., and A. Fausto-Sterling. 2003. "Educating for Social Responsibility: Changing the Syllabus of Developmental Biology." *International Journal of Developmental Biology* 47 (2–3): 327–244.

Gilbert, S. F., and R. Howes-Mischel. 2004. "'Show Me Your Original Face Before You Were Born': The Convergence of Public Fetuses and Sacred DNA." *History and Philosophy of the Life Sciences* 26 (3–4): 377–394.

Gilbert, S. F., A. L. Tyler, and E. J. Zackin. 2005. *Bioethics and the New Embryology: Springboards for Debate*. Sunderland, MA: Sinauer.

Gleicher, N., D. M. Oleske, I. Tur-Kaspa, A. Vidali, and V. Karande. 2000. "Reducing the Risk of High-Order Multiple Pregnancy after Ovarian Stimulation with Gonadotropins." *New England Journal of Medicine* 343 (1): 2–7.

Gluckman P., and M. Hanson. 2004. *The Fetal Matrix: Evolution, Development and Disease*. Cambridge: Cambridge University Press.

——. 2007. *Mismatch: Why Our World No Longer Fits Our Bodies*. Oxford: Oxford University Press.

Goldberg, A. E. 2010. *Lesbian and Gay Parents and Their Children: Research on the Family Life Cycle*. Washington, DC: American Psychological Association.

Goldenberg, S. 2006. "Woman, 62, Gives Birth to Twelfth Child." *Guardian*, February 23. www.theguardian.com/world/2006/feb/23/usa.suzannegoldenberg.

Gould, S. J. 1977. *Ever Since Darwin*. New York: Norton.

———. 1999. *Rocks of Ages: Science and Religion in the Fullness of Life*. New York: Ballantine.

Grady, D. 2017. "Patients Lose Sight After Stem Cells Are Injected into Their Eyes." *New York Times*, March 17. https://www.nytimes.com/2017/03/15/health/eyes-stem-cells -injections.html?_r=0.

Greene, N. D., and A. J. Copp. 2014. "Neural Tube Defects." *Annual Review of Neuroscience* 37: 221–242.

Greenhouse, L., and R. B. Siegel, eds. 2012. *Before Roe v. Wade: Voices that Shaped the Abortion Debate Before the Supreme Court's Ruling*. 2nd ed. New Haven, CT: Yale Law School, Lillian Goldman Law Library. http://documents.law.yale.edu/before-roe.

Griffith, A. J., W. Ji, M. E. Prince, R. A. Altschuler, and M. H. Meisler. 1999. "Optic, Olfactory, and Vestibular Dysmorphogenesis in the Homozygous Mouse Insertional Mutant Tg9257." *Journal of Craniofacial Genetics and Developmental Biology* 19 (3): 157–163.

Grobstein, C. 1988. *Science and the Unborn: Choosing Human Futures*. New York: Basic Books.

Gurdon, J. B. 1962. "The Developmental Capacity of Nuclei Taken from Intestinal Epithelium Cells of Feeding Tadpoles." *Journal of Embryology and Experimental Morphology* 10: 622–640.

Guttmacher, A. F. 1943. "The Role of Artificial Insemination in the Treatment of Human Sterility." *Bulletin of the New York Academy of Medicine* 19: 573–591.

Hanna, E., and B. Gough. 2015. "Experiencing Male Infertility: A Review of the Quantitative Research Literature." *Sage Open* 5 (4): 1–9. doi:10.1177/2158244015610319.

Hanna, J., M. Wernig, S. Markoulaki, C. W. Sun, A. Meissner, J. P. Cassady, C. Beard, T. Brambrink, L. C. Wu, T. M. Townes, and R. Jaenisch. 2007. "Treatment of Sickle Cell Anemia Mouse Model with iPS Cells Generated from Autologous Skin." *Science* 318 (5858): 1920–1923.

Haraway, D. 2015. "Anthropocene, Capitalocene, Plantationocene, Chthulucene: Making Kin." *Environmental Humanities* 6 (1): 159–165.

Haraway, D. J. 2016. *Staying with the Trouble*. Durham, NC: Duke University Press.

Harris, J. 2010. *Enhancing Evolution: The Ethical Case for Making Better People*. Princeton: Princeton University Press.

Harris, J. and Darnovsky, M. 2016. "Pro and Con: Should Gene Editing Be Performed on Human Embryos?" Center for Genetics and Society. http://www.geneticsandsociety .org/article.php?id=9553.

Harris, L. H. 2012. "Recognizing Conscience in Abortion Provision." *New England Journal of Medicine* 367 (11): 981–983.

Heschel, A. J. 1954. *God in Search of Man*. New York: Harper and Row.

———. 1965. *Who Is Man?* Berkeley, CA: University of California Press.

Hesketh, T., L. Lu, and Z. W. Xing. 2011. "The Consequences of Son Preference and Sex-Selective Abortion in China and Other Asian Countries." *Canadian Medical Association Journal* 183 (12): 1374–1377.

Holub, M. 1990. "The Intimate Life of Nude Mice." In M. Holub, *The Dimension of the Present Moment: Essays*, translated by D. Hábová and D. Young. London: Faber and Faber.

Huckabee, M. 2012. "Interview with Jon Stewart." By Jon Stewart. *The Daily Show with Jon Stewart*, November 12. www.cc.com/video-clips/zqfoba/the-daily-show-with -jon-stewart-mike-huckabee-pt—2.

——. 2015. Quoted in B. Guarino, "Mike Huckabee Says Life Begins with a 'DNA Schedule,' a Made-Up Phrase." *Inverse*, August 7. www.inverse.com/article/5174 -mike-huckabee-says-life-begins-with-a-dna-schedule-a-made-up-phrase.

Hudson, V. M., and A. M. den Boer. 2004. *Bare Branches: The Security Implications of Asia's Surplus Male Population*. Cambridge, MA: MIT Press.

Hunt, P. A., C. Lawson, M. Gieske, B. Murdoch, H. Smith, A. Marre, T. Hassold, and C. A. VandeVoort. 2012. "Bisphenol A Alters Early Oogenesis and Follicle Formation in the Fetal Ovary of the Rhesus Monkey." *Proceedings of the National Academy of Sciences of the United States of America* 109 (43): 17525–17530.

Huxley, A. 1937. *Means and Ends*. London: Chatto and Windus.

Huxley, T. H. 1870. "On Descartes' 'Discourse Touching the Method of Using One's Reason Rightly and of Seeking Scientific Truth.'" *Macmillan's Magazine*, March 24.

——. 1894. *Evolution and Ethics: Prolegomena*. London: Macmillan. Available at http:// alepho.clarku.edu/huxley/CE9/E-EProl.html.

Hvistendahl, M. 2011. *Unnatural Selection: Choosing Boys Over Girls, and the Consequences of a World Full of Men*. New York: Public Affairs.

Hyun, I., A. Wilkerson, and J. Johnston. 2016. "Embryology Policy: Revisit the 14-day Rule." *Nature* 533 (7602): 169–171.

Inhorn, M. C. 1994a. *Quest for Conception: Gender, Infertility and Egyptian Medical Traditions*. Philadelphia: University of Pennsylvania Press.

——. 1994b. *Infertility and Egyptian Medical Traditions*. Philadelphia: University of Pennsylvania Press.

——. 1995. *Infertility and Patriarchy: The Cultural Politics of Gender and Family Life in Egypt*. Philadelphia: University of Pennsylvania Press.

——. 2003. *Local Babies, Global Science: Gender, Religion, and In-Vitro Fertilization in Egypt*. New York: Routledge.

——. 2015. *Cosmopolitan Conceptions: IVF Sojourns in Global Dubai*. Durham, NC: Duke University Press.

Inhorn, M. C., and F. van Balen. 2002. "Interpreting Infertility: A View from the Social Sciences." In *Infertility Around the Globe: New Thinking on Childlessness, Gender, and Reproductive Technologies*, edited by M. C. Inhorn and F. van Balen, 3–32. Berkeley: University of California Press.

Jadva, V., and S. Imrie. 2014. "Children of Surrogate Mothers: Psychological Well-Being, Family Relationships and Experiences of Surrogacy." *Human Reproduction* 29 (1): 90–96.

Jaenisch, R., and I. Wilmut. 2001. "Developmental Biology: Don't Clone Humans!" *Science* 291 (5513): 2552.

Jakobovits, I. 1973. "Jewish Views on Abortion." In *Abortion, Society, and Law*, edited by D. Walbert and J. Butler. Cleveland: Press of Case Western Reserve University.

Joffe, C. E. 1995. *Doctors of Conscience: The Struggle to Provide Abortion Before and After Roe v. Wade*. Boston: Beacon.

Johnston, J., and M. Zoll. 2014. "Is Freezing Your Eggs Dangerous? A Primer." *New Republic*, November 1. https://newrepublic.com/article/120077/dangers-and-realities -egg-freezing.

Josephs, L. 2005. "Therapist Anxiety About Motivation for Parenthood." In *Frozen Dreams: Psychodynamic Dimensions of Infertility and Assisted Reproduction*, edited by A. Rosen and J. Rosen, 33–50. Hillsdale, NJ: Analytic.

Kaiser, J., and D. Normile 2015. "Bioethics: Embryo Engineering Study Splits Scientific Community." *Science* 348 (6234): 486–487.

Kalb, C. 2004. "Brave New Babies." *Newsweek*, January 26. At http://www.newsweek .com/brave-new-babies-125951.

Kane, E. 1998. *Birth Mother: The Story of America's First Legal Surrogate Mother*. San Diego: Harcourt Brace Jovanovich.

Kevles, B. 1998. *Naked to the Bone: Medical Imaging in the Twentieth Century*. New York: Basic Books.

Kevles, D. J. 1998. *In the Name of Eugenics: Genetics and the Uses of Human Heredity*. Cambridge, MA: Harvard University Press.

Klein, R. D. 1989. *Infertility: Women Speak Out About Their Experiences of Reproductive Medicine*. London: Pandora.

Knoepfler, P. 2017. "NEJM Paper Links 3 Blinded Patients to Publicly-Traded Stem Cell Clinic." *The Niche* (blog), Knoepfler Lab, March 15. https://ipscell.com/2017/03 /nejm-paper-links-3-patients-blinded-to-publicly-traded-stem-cell-clinic/.

Kohlberg, K. 1953a. "The Practice of Artificial Insemination in Humans." *Deutsche medizinische Wochenschrift* 78: 835–839.

——. 1953b. "Artificial Insemination and the Physician." *Deutsche medizinische Wochenschrift*. 78: 855–856.

Kolbert, E. 2014. *The Sixth Extinction: An Unnatural History*. New York: Henry Holt.

Kühl, S. 1994, *The Nazi Connection: Eugenics, American Racism, and German National Socialism*. New York: Oxford University Press.

Kuriyan, A. E., T. A. Albini, J. H. Townsend, et al. 2017. "Vision Loss After Intravitreal Injection of Autologous 'Stem Cells' for AMD." *New England Journal of Medicine* 376: 1047–1053.

Lakoff, G., and M. Johnson. 1980. *Metaphors We Live By*. Chicago: University of Chicago Press.

Lambert, N. 2015. "A Modern Woman's Burden." *New Republic*, March 20. https:// newrepublic.com/article/121334/modern-womans-burden.

Lathi, R. B., C. A. Liebert, K. F. Brookfield, J. A. Taylor, F. S. vom Saal, V. Y. Fujimoto, and V. L. Baker. 2014. "Conjugated Bisphenol A in Maternal Serum in Relation to Miscarriage Risk." *Fertility and Sterility* 102 (1): 123–128.

Leder A., P. K. Pattengale, A. Kuo, T. A. Stewart, and P. Leder. 1986. "Consequences of Widespread Deregulation of the *c-myc* Gene in Transgenic Mice: Multiple Neoplasms and Normal Development." *Cell* 45 (4): 485–495.

Leigh, J. 2016. *Avalanche: A Love Story*. New York: Norton.

Lemos, E. V., D. Zhang, B. J. Van Voorhis, and X. H. Hu. 2013. "Healthcare Expenses Associated with Multiple vs Singleton Pregnancies in the United States." *American Journal of Obstetrics and Gynecology* 209 (6): 586.e1–586.e11.

Levin, I. 1976. *The Boys from Brazil*. New York: Random House.

Li, H., L. Saucedo-Cuevas, J. A. Regla-Nava, G. Chai, N. Sheets, W. Tang, A. V. Terskikh, S. Shresta, and J. G. Gleeson. 2016. "Zika Virus Infects Neural Progenitors in the Adult Mouse Brain and Alters Proliferation." *Cell Stem Cell* 19 (5): 593–598.

Liang, P., Y. Xu, X. Zhang, C. Ding, R. Huang, Z. Zhang, J. Lv, X. Xie, Y. Chen, Y. Li, Y. Sun, Y. Bai, Z. Songyang, W. Ma, C. Zhou, and J. Huang. 2015. "CRISPR/Cas9 -Mediated Gene Editing in Human Tripronuclear Zygotes." *Protein Cell* 6 (5): 363–372.

Lipshultz, L., and D. Adamson. 1999. "Multiple-Birth Risk Associated with In Vitro Fertilization: Revised Guidelines." *Journal of the American Medical Association* 282 (19): 1813–1814.

Loades, D. M. 2006. *Mary Tudor: The Tragical History of the First Queen of England*. Kew, UK: National Archives.

Lorenceau, E., L. Mazzucca, S. Tisseron, and T. D. Pizitz. 2015. "A Cross-Cultural Study on Surrogate Mother's Empathy and Maternal-foetal Attachment. "*Women and Birth* 28 (2): 154–159.

Ludmerer, K. M. 1972. *Genetics and American Society: A Historical Appraisal*. Baltimore: Johns Hopkins University Press.

Lunneborg, P. 2002. *The Chosen Lives of Childfree Men*. London: Bergin and Garvey.

Macklin, R. 1995. "The Ethics of Sex Selection." *Indian Journal of Medical Ethics* 3: 61–64.

Maimonides (Moshe ben Maimon). (1190) 1956. *The Guide for the Perplexed*, translated by M. Friedlander. New York: Dover.

Mantzouratou, A., and J. D. Delhanty. 2011. "Aneuploidy in the Human Cleavage Stage Embryo." *Cytogenetic and Genome Research* 133 (2–4): 141–148.

Marsh, M., and W. Ronner. 1996. *The Empty Cradle: Infertility in America from Colonial Times to the Present*. Baltimore: Johns Hopkins University Press.

Martin, E. 1991. "The Egg and the Sperm: How Science Has Constructed a Romance Based on Stereotypical Male–Female Roles." *Signs* 16 (3): 485–501.

Martin, G. R. 1981. "Isolation of a Pluripotent Cell Line from Early Mouse Embryos Cultured in Medium Conditioned by Teratocarcinoma Stem Cells." *Proceedings of the National Academy of Sciences of the United States of America* 78 (12): 7634–7638.

Mason, M.-C. 1993. *Male Infertility: Men Talking*. New York: Routledge.

Mayo Clinic. 2014. "Ovarian Hyperstimulation Syndrome." www.mayoclinic.org /diseases-conditions/ovarian-hyperstimulation-syndrome-ohss/basics/definition /con-20033777.

McConnell, D. 2010. "Cristiano Ronaldo Paid a Surrogate Mother to Have His Baby." *Daily Mail*. http://www.dailymail.co.uk/tvshowbiz/article-1292094/Cristiano-Ronaldo -father-paying-surrogate-baby.html.

McCormick, R. 1991. "Who or What Is the Pre-Embryo?" *Kennedy Institute of Ethics Journal* 1 (1): 1–15.

McDonald, J. W., X. Z. Liu, Y. Qu, S. Liu, S. K. Mickey, D. Turetsky, D. I. Gottlieb, and D. W. Choi. 1999. "Transplanted Embryonic Stem Cells Survive, Differentiate and Promote Recovery in Injured Rat Spinal Cord." *Nature Medicine* 5 (12): 1410–1412.

McLaren, A. 2007. *Impotence: A Cultural History*. Chicago: University of Chicago Press.

McPherron, A. C., A. M. Lawler, and S. J. Lee. 1997. "Regulation of Skeletal Muscle Mass in Mice by a New TGF-Beta Superfamily Member." *Nature* 387 (6628): 83–90.

Meade, H. M. 1997. "Dairy Gene." *The Sciences* 37 (5): 20–25.

Meaney, M. J. 2001. "Maternal Care, Gene Expression, and the Transmission of Individual Differences in Stress Reactivity Across Generations." *Annual Review of Neuroscience* 24: 1161–1192.

Mei, G. 2016. "Italy Tells Women to Hurry Up and Have Babies." *Redbook*, September 2. www.redbookmag.com/body/pregnancy-fertility/a45777/italy-fertility -day-offensive-ads/.

Melo, E. O., A. M. Canavessi, M. M. Franco, and R. Rumpf. 2007. "Animal Transgenesis: State of the Art and Applications." *Journal of Applied Genetics* 48 (1): 47–61.

Mitchell, C., L. M. Schneper, and D. A. Notterman. 2016. "DNA Methylation, Early Life Environment, and Health Outcomes." *Pediatric Research* 79 (1–2): 212–219.

Montagu, M. F. A. 1962. "Time, Morphology, and Neoteny in the Evolution of Man." In *Culture and Evolution of Man*, edited by M. F. A. Montagu, 324–342. New York: Oxford University Press.

Morowitz, H. J., and J. S. Trefil. 1992. *The Facts of Life: Science and the Abortion Controversy*. New York: Oxford University Press.

Mullin, E. 2017. "Eggs from Skin Cells? Here's Why the Next Fertility Technology Will Open Pandora's Box." *MIT Technology Review*. https://www.technologyreview .com/s/603343/eggs-from-skin-cells-heres-why-the-next-fertility-technology-will -open-pandoras-box/.

National Library of Medicine (NLM). 2016. "In Vitro Fertilization." *MedlinePlus*. www .nlm.nih.gov/medlineplus/ency/article/007279.htm.

Needham, J. 1931. *Chemical Embryology*. New York: MacMillan.

Needham, J. (1934) 1959. *A History of Embryology*. Cambridge: Cambridge University Press.

Nelkin, D., and M. S. Lindee. (1995) 2004. *The DNA Mystique: The Gene as a Cultural Icon*. Ann Arbor: University of Michigan Press.

Nelson, C. 2014. "What a Wrongful Birth Lawsuit Can Teach Us About Race." *Huffington Post,* October 7. www.huffingtonpost.ca/charmaine-nelson/wrongful-birth -ohio_b_5946982.html.

New Family. 2016. http://awiderbridge.org/new-family/

New York Blood Center (NYBC). "NY Blood Center: National Cord Blood Program." http://parentsguidecordblood.org/en/banks/ny-blood-center-national-cord-blood -program.

New York State Department of Health (NYSDH). 2103. "Cord Blood Frequently Asked Questions." www.health.ny.gov/professionals/patients/donation/umbilical_cord_blood /frequently_asked_questions.htm.

New York Times. 1992. "Doctor Is Found Guilty in Fertility Case." March 5. www.nytimes .com/1992/03/05/us/doctor-is-found-guilty-in-fertility-case.html.

Newman, S. A. 2003. "Averting the Clone Age: Prospects and Perils of Human Developmental Manipulation. *Journal of Contemporary Health Law and Policy.* 19 (2): 431–463.

——. 2013. "The British Embryo Authority and the Chamber of Eugenics." *Huffington Post.* http://www.huffingtonpost.com/stuart-a-newman/mitochondrial-replacement -ethics_b_2837818.html.

Nilsson, L. 1965. "Drama of Life Before Birth." *Life,* April 30, 54–72A.

——. 1966. *A Child Is Born: The Drama of Life Before Birth.* New York: Dell.

Noé, G., H. B. Croxatto, A. M. Salvatierra, V. Reyes, C. Villarroel, C. Muñoz, G. Morales, and A. Retamales. 2011. "Contraceptive Efficacy of Emergency Contraception with Levonorgestrel Given Before or After Ovulation." *Contraception* 84 (5): 486–492.

Nosarka, S., and T. F. Kruger. 2005. "Surrogate Motherhood." *South African Medical Journal* 95 (12): 942, 944, 946.

Ombelet, W., and J. Van Robays. 2015. "Artificial Insemination: Hurdles and Milestones." *Facts, Views & Vision in ObGyn* 7 (2): 137–143.

Pagliuca, F. W., J. R. Millman, M. Gürtler, M. Segel, A. Van Dervort, J. H. Ryu, Q. P. Peterson, D. Greiner, and D. A. Melton. 2014. "Generation of Functional Human Pancreatic β Cells in Vitro." *Cell* 159 (2): 428–439.

Pande, A. 2014. *Wombs in Labor: Transnational Commercial Surrogacy in India.* New York: Columbia University Press.

Papaligoura, Z., D. Papadatou, and T. Bellali. 2015. "Surrogacy: The Experience of Greek Commissioning Women." *Women and Birth* 28 (4): e110–e118.

Petersen, K. B., H. W. Hvidman, R. Sylvest, A. Pinborg, E. C. Larsen, K. T. Macklon, A. N. Andersen, and L. Schmidt. 2015. "Family Intentions and Personal Considerations on Postponing Childbearing in Childless Cohabiting and Single Women Age 35–43 Seeking Fertility Assessment and Counseling." *Human Reproduction* 30 (11): 2563–2574.

Peterson, R. 2003. *Scenes from a Surrogacy.* Hillsdale, NJ: Analytic.

Pinto-Correia, C. 1986. *O Essencial Sobre os Bebes-Proveta.* Lisbon: Imprensa Nacional— Casa da Moeda.

——. 1997. *The Ovary of Eve: Sperm & Egg & Preformation*. Chicago: University of Chicago Press.

——. 2003. *Return of the Crazy Bird: The Strange Tale of the Dodo*. New York: Copernicus.

Pius IX. 1869. *Apostolicae Sedis*.

Pollack, A. 2009. "FDA Approves Drug from Gene-Altered Goats." *New York Times*, February 6. www.nytimes.com/2009/02/07/business/07goatdrug.html?pagewanted =all&_r=0.

Practice Committees of the American Society for Reproductive Medicine and the Society for Assisted Reproductive Technology. 2013. "Mature Oocyte Cryopreservation: A Guideline." *Fertility and Sterility* 99 (1): 37–43.

Presbyterian Mission Agency (PMA). 1992. http://www.presbyterianmission.org/what -we-believe/social-issues/abortion-issues/

Pro-Life Action League. 2003. http://www.prolifeaction.org/actionnewshotline/2003 /0604-0608.htm.

Purdy, L. 2001/2015. "Bioethics of New Assisted Reproduction." *Encyclopedia of Life Sciences*. http://onlinelibrary.wiley.com/doi/10.1002/9780470015902.a0003479.pub3 /full

Purves, D., and J. W. Lichtman. 1985. *Principles of Neural Development*. Sunderland, MA: Sinauer.

Rabinowitz, A. 2015. "Why Egg Freezing Is an Impossible Choice." *Nautilus* 22, March 19. http://nautil.us/issue/22/slow/why-egg-freezing-is-an-impossible-choice.

Raff, R. A. 2012. *Once We All Had Gills: Growing Up Evolutionist in an Evolving World*. Bloomington: Indiana University Press.

Ramachandran, R. 1999. "In India, Sex Selection Gets Easier." *UNESCO Courier*. http:// www.unesco.org/courier/1999_09/uk/dossier/txt06.htm.

Ramsey, P. 1970. "Reference Points in Deciding About Abortion." In *The Morality of Abortion: Legal and Historical Perspectives*, edited by J. T. Noonan, 40–100. Cambridge, MA: Harvard University Press.

Renfree, M. B. 1982. "Implantation and Placentation." In *Reproduction in Mammals. Book 2. Embryonic and Fetal Development*, edited by C. R. Austin and R. V. Short. 2nd ed. Cambridge: Cambridge University Press.

Resolve. 2016. "What Are My Chances of Success with IVF?" www.resolve.org/family -building-options/ivf-art/what-are-my-chances-of-success-with-ivf.html.

Rezania, A., J. E. Bruin, P. Arora, A., Rubin, I. Batushansky, A. Asadi, S. O'Dwyer, N. Quiskamp, M. Mojibian, T. Albrecht, Y. H. Yang, J. D. Johnson, and T. J. Kieffer. 2014. "Reversal of Diabetes with Insulin-Producing Cells Derived in Vitro from Human Pluripotent Stem Cells." *Nature Biotechnology* 32 (11): 1121–1133.

Rifkin, J. 1998. *The Biotech Century*. New York: Putnam.

Roberts, J. C. 2002. "Customizing Conception: A Survey of Preimplantation Genetic Diagnosis and the Resulting Social, Ethical, and Legal Dilemmas." *Duke Law and Technology Review*, http://www.law.duke.edu.edu/journals/dltr/articles/2002dltr0012 .html.

Robertson, J. A. 2001. "Preconception Gender Selection." *The American Journal of Bioethics* 1 (1): 2–5.

Robinton, D. A., and G. Q. Daley. 2012. "The Promise of Induced Pluripotent Stem Cells in Research and Therapy." *Nature* 481 (7381): 295–305.

Rorvik, D. 1978. *In His Image: The Cloning of a Man*. Philadelphia: Lippincott.

Roura, S., J. M. Pjual, C. Gálvez-Montón, and A. Bayes-Genis. 2016. "Quality and Exploitation of Umbilical Cord Blood for Cell Therapy: Are We Beyond Our Capabilities?" *Developmental Dynamics* 245 (7): 710–717.

Rose, S. 1998. *Lifelines: Biology Beyond Determinism*. Oxford: Oxford University Press.

Rosen, A., and J. Rosen, eds. 2005. *Frozen Dreams: Psychodynamic Dimensions of Infertility and Assisted Reproduction*. Hillsdale, NJ: Analytic.

Rostand, J. 1962. *The Substance of Man*, translated by I. Brandeis. New York: Doubleday.

Roy, R. N., W. R. Schumm, and S. L. Britt. 2014. *Transition to Parenthood*. New York: Springer.

Sandel, M. J. 2004. "Ethics of Embryonic Stem Cells." *New England Journal of Medicine* 351 (16): 1687–1690.

——. 2005. *Public Philosophy: Essays on the Morality in Politics*. Cambridge, MA: Harvard University Press.

Sander, J. D., and J. K. Joung. 2014. "CRISPR-Cas Systems for Editing, Regulating and Targeting Genomes." *Nature Biotechnology* 32 (4): 347–355.

Satpathy, R. and S. K. Mishra. 2000. "The Alarming 'Gender Gap.'" *Bulletin of the World Health Organization* 78 (11): 1373.

Satouh Y., N. Inoue, M. Ikawa, and M. Okabe. 2012. "Visualization of the Moment of Mouse Sperm–Egg Fusion and Dynamic Localization of IZUMO1." *Journal of Cell Science* 125 (Part 21): 4985–4990.

Schatten, G., and H. Schatten. 1983. "The Energetic Egg." *The Sciences* 23 (5): 28–34.

Schieve, L. A., H. B. Peterson, S. F. Meikle, G. Jeng, I. Danel, N. M. Burnett, and L. S. Wilcox. 1999. "Live-Birth Rates and Multiple-Birth Risk Using In Vitro Fertilization." *Journal of the American Medical Association* 282 (19): 1832–1838.

Schnieke, A. E., A. J. Kind, W. A. Ritchie, K. Mycock, A. R. Scott, M. Ritchie, I. Wilmut, A. Colman, and K. H. Campbell. 1997. "Human Factor IX Transgenic Sheep Produced by Transfer of Nuclei from Transfected Fetal Fibroblasts." *Science* 278 (5346): 2130–2134.

Schuelke, M., K. R. Wagner, L. E. Stolz, C. Huber, T. Riebel, W. Komen, T. Braun, J. F. Tobin, S.-J. Lee. 2004. "Myostatin Mutation Associated with Gross Muscle Hypertrophy in a Child." *New England Journal of Medicine* 350: 2682–2688.

Schwartz, B. 2004. *The Paradox of Choice*. New York: Harper Perennial.

Sclaff, W. D., and A. M. Braverman. 2015. "Mental Health Counseling in Third-Party Reproduction in the United States: Evaluation, Psychoeducation, or Ethical Gatekeeping?" *Fertility and Sterility* 104 (3): 501–506.

Settle, J. E., C. T. Dawes, N. A. Christakis, and J. H. Fowler. 2010. "Friendships Moderate an Association Between a Dopamine Gene Variant and Political Ideology." *Journal of Politics* 72 (4): 1189–1198.

Shane, M., and L. Wilson. 2013. *After Tiller*. DVD. Directed by M. Shane and L. Wilson. New York: Oscilloscope Laboratories.

Shannon, T. A., and A. B. Wolter. 1990. "Reflections on the Moral Status of the Pre-Embryo." *Theological Studies* 51 (4): 603–626.

Shapiro, G. K. 2014. "Abortion Law in Muslim-Majority Countries: An Overview of the Islamic Discourse with Policy Implications." *Health Policy and Planning* 29 (4): 483–494.

Shete, M. 2005. "Doc in the Dock." *Times of India*, April 15.

Sibov, T. T., P. Severino, L. C. Marti, L. F. Pavon, D. M. Oliveira, P. R. Tobo, A. H. Campos, A. T. Paes, E. Amaro Jr., L. F. Gamarra, and C. A. Moreira-Filho. 2012. "Mesenchymal Stem Cells from Umbilical Cord: Parameters for Isolation, Characterization, and Adipogenic Differentiation." *Cytotechnology* 64 (5): 511–521.

Silver, L. 1998. *Remaking Eden: How Genetic Engineering and Cloning Will Transform the American Family*. New York: Avon.

Skloot, R. 2010. *The Immortal Life of Henrietta Lacks*. New York: Crown.

Small, M. 1991. "Sperm Wars." *Discover*, July, 48–53.

Smith, R. 2012. "British Man 'Fathered 600 Children' at Own Fertility Clinic." *Telegraph*, April 8. www.telegraph.co.uk/news/9193014/British-man-fathered-600-children-at -own-fertility-clinic.html.

Sneed, A. 2014. "Fact or Fiction? Emergency Contraceptives Cause Abortions." *Scientific American*. https://www.scientificamerican.com/article/fact-or-fiction-emergency -contraceptives-cause-abortions/

Southern Baptist Convention. 1999. "Southern Baptist Convention Resolutions on Abortion." www.johnstonsarchive.net/baptist/sbcabres.html.

Spangmose, A. L., S. S. Malchau, L. Schmidt L, et al. 2017. "Academic Performance in Adolescents Born After ART—a Nationwide Registry-Based Cohort Study." *Human Reproduction*. doi: 10.1093/humrep/dew334.

Spar, D. 2006. *The Baby Business: How Money, Science, and Politics Drive the Commerce of Conception*. Boston: Harvard Business Review Press.

Speroff, L., and M. A. Fritz. 2005. *Clinical Gynecologic Endocrinology and Infertility*. 7th ed. Philadelphia: Lippincott Williams and Wilkins.

Steiner, L. M. 2013. *The Baby Chase: How Surrogacy Is Transforming the American Family*. New York: St. Martin's.

Stock, G. 2002. *Redesigning Humans: Our Inevitable Genetic Future*. Boston: Houghton Mifflin.

Stoughton, R. H. 1948. "Artificial Human Insemination." *Nature* 13: 790.

Takahashi, K., K. Tanabe, M. Ohnuki, M. Narita, T. Ichisaka, K. Tomoda, and S. Yamanaka. 2007. "Induction of Pluripotent Stem Cells from Adult Human Fibroblasts by Defined Factors." *Cell* 131 (5): 861–872.

Takahashi, K., and S. Yamanaka. 2006. "Induction of Pluripotent Stem Cells from Mouse Embryonic and Adult Fibroblast Cultures by Defined Factors." *Cell* 126 (4): 663–676.

Tang, B.L. 2016. "Zika Virus as a Causative Agent for Primary Microencephaly: the Evidence so Far." *Archives of Microbiology* 198 (7): 595–601.

Tarkowski, A. K., A. Suwińska, R. Czołowska, and W. Ożdżeński. 2010. "Individual Blastomeres of 16- and 32-Cell Mouse Embryos Are Able to Develop into Foetuses and Mice." *Developmental Biology* 348 (2): 190–198.

Tertullian. A.D. 197. *Apologia*, chapter 9. Translated by Rev. S. Thelwall. www.earlychristian writings.com/text/tertullian01.html.

TheNotMom.com. 2016. "Childless Women Stories." http://thenotmom.com/tag/childless -women-stories/.

Thomas, J. A., and V. Samson. 2013. *The Baby Game*. (Self-published).

Throsby, K. 2004. *When IVF Fails: Feminism, Infertility, and the Negotiation of Normality*. London: Palgrave Macmillan.

Throsby, K., and R. Gill. 2004. " 'It's Different for Men': Masculinity and IVF." *Men and Masculinities* 6 (4): 330–348.

Tillich, P. 1952. *Dynamics of Faith*. New York: Harper and Row.

Tribe, L. 1990. *Abortion: The Clash of the Absolutes*. New York: Norton.

Trotsky, L. 1935. "If America Should Go Communist." *Liberty*, March 23. www.marxists .org/archive/trotsky/1934/08/ame.htm.

Trounson, A. O., and D. K. Gardner, eds. 2000. *Handbook of In Vitro Fertilization*. 2nd ed. Boca Raton, FL: CRC.

Uffalussy, J. G. 2014. "The Cost of IVF: Four Things I Learned While Battling Infertility." *Forbes Magazine*, February 6. www.forbes.com/sites/learnvest/2014/02/06/the-cost-of -ivf-4-things-i-learned-while-battling-infertility/.

Urban, M. C. 2015. "Climate Change: Accelerating Extinction Risk from Climate Change." *Science* 348 (6234): 571–573.

Van Heesch, M. M., J. L. H. Evers, M. A. H. B. M. van der Hoeven, J. C. M. Dumoulin, C. E. M. van Beijsterveldt, G. J. Bonsel, R. H. M. Dykgraaf, J. B. van Goudoever, C. Koopman-Esseboom, W. L. D. M. Nelen, K. Steiner, P. Tamminga, N. Tonch, H. L. Torrance, and C. D. Dirksen. 2015. "Hospital Costs During the First Five Years of Life for Multiples Compared with Singletons Born after IVF or ICSI." *Human Reproduction* 30 (6): 1481–1490.

Van Niekerk, A., and L. van Zyl, 1995. "The Ethics of Surrogacy: Women's Reproductive Labour." *Journal of Medical Ethics* 21 (6): 345–349.

Vargas, M. F., A. A. Tapia-Pizarro, S. P. Henríquez, M. Quezada, A. M. Salvatierra, G. Noe, D. J. Munroe, L. A. Velasquez, and H. B. Croxatto. 2012. "Effect of Single Post-ovulatory Administration of Levonorgestrel on Gene Expression Profile During the Receptive Period of the Human Endometrium." *Journal of Molecular Endocrinology* 48 (1): 25–36.

Vikström, J., G. Sydsjö, M. Hammar, M. Bladh, and A. Josefsson. 2015. "Risk of Postnatal Depression or Suicide After in Vitro Fertilisation Treatment: A Nationwide Case–Control Study." *British Journal of Obstetrics and Gynaecology*. doi:10.1111/1471-0528.13788.

Vines, G. 1993. "The Hidden Cost of Sex Selection." *New Scientist*, May 1, 12–13.

Vissing, Y. 2002. *Women Without Children: Nurturing Lives*. Rutgers: Rutgers University Press.

Volpe, E. P. 1987. *Test-Tube Conception: A Blend of Love and Science*. Macon, GA: Mercer University Press.

Wagner, G. 1981. *Impotence: Physiological, Psychological, and Surgical Diagnosis and Treatment*. New York: Springer.

Wang, H., and S. K. Dey. 2006. "Roadmap to Embryo Implantation: Clues from Mouse Models." *Nature Reviews Genetics* 7 (3): 185–199.

Washington, G. 2014. "Genius Sperm." *Snap Judgment*. National Public Radio. October 3. www.npr.org/2014/10/03/353491991/genius-sperm.

Watson, J. 2000. "The Road Ahead," In *Engineering the Human Germline*, edited by G. Stock and J. Campbell. Oxford: Oxford University Press.

——. 2016. "Eugenics and Bioethics: Interview with James Watson." By Cold Spring Harbor Laboratory. *DNA Learning Center*. www.dnalc.org/view/15472-Eugenics-and-bioethics-James-Watson.html.

Watson, J. E., D. F. Shanahan, M. Di Marco, J. Allan, W. F. Laurance, E. W. Sanderson, B. Mackey, and O. Venter. 2016 "Catastrophic Declines in Wilderness Areas Undermine Global Conservation Targets." *Current Biology* 26 (21): 1–6.

Weiss, R. 1998. "Science on the Ethical Frontier: Engineering the Unborn." *Washington Post*, March 22. www.washingtonpost.com/wp-srv/national/science/ethical/unborn.htm.

White, A. D. (1896) 1960. *A History of the Warfare of Science with Theology in Christendom*. New York: Dover.

Whitehead, M. B., and L. Schwartz-Nobel. 1989. *A Mother's Story: The Truth about the Baby M Case*. New York: St. Martin's.

Wilmut, I. In Wilmut, I., K. Campbell, and C. Trudge. 2000. *The Second Creation: Dolly and the Age of Biological Control*. New York: Farrar, Straus and Giroux.

Wilmut, I., A. E. Schnieke, J. McWhir, A. J. Kind, and K. H. Campbell. 1997. "Viable Offspring Derived from Fetal and Adult Mammalian Cells." *Nature* 385 (6619): 810–814.

Wilson, E. O. 1980. *Sociobiology*. Cambridge, MA: Belknap.

Wilson, J. M. 2009. "Lessons Learned from the Gene Therapy Trial for Ornithine Transcarbamylase Deficiency." *Molecular and Genetic Metabolism* 96 (4): 151–157.

Wilson, K. J. 2014. *Not Trying: Infertility, Childlessness, and Ambivalence*. Nashville: Vanderbilt University Press.

Winkvist, A. and H. Z. Ahktar. 2000. "God Should Give Daughters to Rich Families Only: Attitudes Towards Childbearing Among Low-Income Women in Punjab, Pakistan." *Social Science and Medicine* 51 (1): 73–82.

Witkin, G., A. Tran, J. A. Lee et al. 2013. "What Makes a Woman Freeze: The Impetus Behind Patients' Desires to Undergo Elective Oocyte Cryopreservation." *Fertility and Sterility* 100 (3): S24.

Wolpert, L. 1983. Quoted in J. M. W. Slack, *From Egg to Embryo: Determinative Events in Early Development*. Cambridge: Cambridge University Press.

WWF (World Wildlife Fund). 2016. *Living Planet Report 2016*. "Risk and resilience in a new era." Gland, Switzerland: WWF International.

Yu, J., M. A. Vodyanik, K. Smuga-Otto, J. Antosiewicz-Bourget, J. L. Frane, S. Tian, J. Nie, G. A. Jonsdottir, V. Ruotti, R. Stewart, I. I. Slukvin, and J. A. Thomson. 2007. "Induced Pluripotent Stem Cell Lines Derived from Human Somatic Cells." *Science* 318 (5858): 1917–1920.

Yuko, E. 2016. "The First Artificial Insemination Was an Ethical Nightmare." *Atlantic*, January 8. www.theatlantic.com/health/archive/2016/01/first-artificial-insemination /423198/.

Zhu, W. X., L. Lu, and T. Hesketh. 2009. "China's Excess Males, Sex-Selective Abortion, and One-Child Policy: Analysis of Data from 2005 National Intercensus Survey." *British Medical Journal* 338: b1211.

Zilman, C. 2016. "Why Italy's New 'Fertility Day' Campaign Is a Sexist Mess." *Fortune*, September 2. http://fortune.com/2016/09/02/italy-fertility-day-birthrate-sexism/.

Zolbrod, A. P. 1993. *Men, Women, and Infertility: Intervention and Treatment Strategies*. New York: Lexington.

Zoll, M. 2013. *Cracked Open: Liberty, Fertility, and the Pursuit of High Tech Babies*. Northampton, MA: Interlink.

Zoll, M. 2014. "What Apple and Facebook Don't Know about Egg Freezing." *To the Contrary* (blog). www.pbs.org/to-the-contrary/blog/3692/what-apple-and -facebook-don%E2%80%99t-know-about-egg-freezing.

Zornberg, A. G. 1995. *The Beginning of Desire: Reflections on Genesis*. New York: Doubleday.

Zou, Q, X. Wang, Y. Liu et al. 2015. "Generation of Gene-Target Dogs Using the CRISPR/ Cas9 System." *Journal of Molecualar Cell Biology*. 7 (6): 580–583.

Zouves, C., and J. Sullivan. 1999. *Expecting Miracles: On the Path from Infertility to Parenthood*. New York: Henry Holt.

INDEX